Compulsory Property Acquisition for Urban Densification

Densification has been a central method of achieving smart, sustainable cities across the world. This book explores international examples of the property rights tensions involved in attempting to develop denser, more sustainable cities through compulsory acquisition of property. The case studies from Europe, North and South America, eastern Asia and Australia show how well, or not, property rights have been recognised in each country. Chapters explore the significance of local legal frameworks and institutions in accommodating property rights in the densification process. In particular, the case studies address the following issues and more:

- Whether compulsory acquisition to increase densification is justified in practice and in theory.
- The specific public benefits given for compulsory acquisition.
- The role the development industry plays in facilitating, encouraging or promoting compulsory acquisition.
- What compensation or offsets are offered for acquisition, and how are they funded?
- Is there a local or national history of compulsory property acquisition by government for a range of purposes?
- Is compulsory acquisition restricted to certain types or locations of densification?
- Where existing housing is acquired, are there obligations to provide alternative housing arrangements?

The central aim of the book is to summarise international experiences of the extent to which property rights have or have not been protected in the use of compulsory property acquisition to achieve sustainable cities via urban densification. It is essential reading for all those interested in planning law, property rights, environmental law, urban studies, sustainable urban development and land use policy.

Glen Searle is Honorary Associate Professor of Urban and Regional Planning and Policy at the University of Sydney, and Honorary Associate Professor in Planning, University of Queensland, Australia. He is the former Editor-in-Chief of *Urban Policy and Research.*

Routledge Complex Real Property Rights Series

Series editor: Professor Spike Boydell

University of Technology, Sydney, Australia

Real Property Rights are central to the global economy and provide a legal framework for how society (be it developed or customary) relates to land and buildings. We need to better understand property rights to ensure sustainable societies, careful use of limited resources and sound ecological stewardship of our land and water. Contemporary property rights theory is dynamic and needs to engage thinkers who are prepared to think outside their disciplinary limitations.

The Routledge Complex Real Property Rights Series strives to take a transdisciplinary approach to understanding property rights and specifically encourages heterodox thinking. Through rich international case studies our goal is to build models to connect theory to observed reality, allowing us to inform potential policy outcomes. This series is both an ideal forum and reference for students and scholars of property rights and land issues.

Beyond Communal and Individual Ownership
Indigenous Land Reform in Australia
Leon Terrill

Strata Title Property Rights
Private Governance of Multi-owned Properties
Cathy Sherry

Property Rights and Climate Change
Land-use Under Changing Environmental Conditions
Fennie van Straalen, Thomas Hartmann and John Sheehan

Property, Place and Piracy
Edited by James Arvanitakis and Martin Fredriksson

Wicked Valuations
People and Landed Property
Michael McDermott

Compulsory Property Acquisition for Urban Densification
Edited by Glen Searle

Compulsory Property Acquisition for Urban Densification

Edited by Glen Searle

Routledge
Taylor & Francis Group
LONDON AND NEW YORK

First published 2018
by Routledge
2 Park Square, Milton Park, Abingdon, Oxon OX14 4RN

and by Routledge
605 Third Avenue, New York, NY 10017

First issued in paperback 2021

Routledge is an imprint of the Taylor & Francis Group, an informa business

Publisher's Note
The publisher has gone to great lengths to ensure the quality of this reprint but points out that some imperfections in the original copies may be apparent.

British Library Cataloguing-in-Publication Data
A catalogue record for this book is available from the British Library

Library of Congress Cataloging-in-Publication Data
Names: Searle, Glen, editor.
Title: Compulsory property acquisition for urban densification / edited by Glen Searle.
Description: Abingdon, Oxon [UK] ; New York, NY : Routledge, 2018. | Series: Routledge complex real property rights series
Identifiers: LCCN 2018008589 | ISBN 9781138500990 (hardback : alk. paper) | ISBN 9781315144085 (ebook)
Subjects: LCSH: Eminent domain (International law) | Government purchasing of real property–Law and legislation. | City planning and redevelopment law. | Urbanization. | Right of property. | Compensation (Law)
Classification: LCC K3511 .C66 2018 | DDC 343/.0252–dc23
LC record available at https://lccn.loc.gov/2018008589

Typeset in Goudy
by Wearset Ltd, Boldon, Tyne and Wear

ISBN 13: 978-1-03-209508-0 (pbk)
ISBN 13: 978-1-138-50099-0 (hbk)

Contents

Contributors

Juliane Albrecht is a senior scientist and project manager at the Leibniz Institute of Ecological Urban and Regional Development (IOER) in Dresden, Germany. She holds a PhD in Law and has extensive experience in environmental law and planning law.

Long Cheng is a PhD candidate at the School of Earth and Environmental Sciences, University of Queensland, Australia. Long's research focuses on urban-rural development and land use policies in China. His chapter is part of his PhD research entitled 'Coordinating urban and rural development under contemporary China's land use policy – the Link Policy'.

Alice Christudason is an associate professor at the Department of Real Estate, National University of Singapore. She obtained her PhD in Land Law from King's College London. Alice consults and researches on various aspects of real property law. She is the lead contributor to *Halsbury's Laws of Singapore (Land, Volume 14)* as well as author and co-author of other reference texts on property law and research papers in various international refereed journals. She has also been involved in representations for law reform before Parliamentary Select Committees. Alice is presently on the Valuation Review Board, Ministry of Finance (Singapore) and on the panel of mediators at the Consumers Association of Singapore.

Gary Cox is an independent public policy and urban regeneration consultant. He has worked extensively in Australia and the UK, mostly recently on an estate renewal project in Wandsworth in south-west London. He is also Adjunct Associate Professor in Urban and Regional Planning at the University of Sydney, Australia.

Laura Crommelin is a research fellow and lecturer at City Futures Research Centre, Faculty of Built Environment at UNSW Sydney, undertaking research on high-density living, urban policy-making and post-industrial revitalisation. Laura holds a PhD in urban planning from UNSW and a BA/LLB (Hons) from the University of Melbourne.

Sébastien Darchen is a lecturer in Planning at the School of Earth and Environmental Sciences at the University of Queensland, Australia. His areas of research include urban regeneration, study of the provision of the built environment, and the creative industries.

Hazel Easthope is an associate professor and UNSW Scientia Fellow at City Futures Research Centre, Faculty of Built Environment at UNSW Sydney. With a background in sociology and human geography, her research interests include the intersections between mobility, identity and home; the governance and planning implications of private apartment developments; and the lived experiences of apartment residents.

Charlie Glinka is an active member of the Australian property industry, providing development management consultancy and guidance across a wide range of asset classes. Charlie holds an Honours degree in Property Economics and a Master's degree in Property Development from the University of Technology Sydney, and is currently working on applying artificial intelligence to the development feasibility process.

Thomas Hartmann is an associate professor at Wageningen University & Research, the Netherlands. He teaches and researches in spatial planning on justice and the city, land and water governance, and he has a special expertise in land policies and planning instruments. He is the vice president of the international academic association on Planning, Law and Property Rights (www.plpr-association.org).

Sanne Holtslag-Broekhof is Process Manager of Spatial Planning at the Dutch Kadaster. She regularly publishes on land policy and land management. She studied spatial planning at Wageningen University, where she completed her PhD (2016) titled 'Dealing with private property for public purposes'.

Marta Lora-Tamayo Vallvé is Administrative Law Associate Professor at the Universidad Nacional de Educación a Distancia (UNED), Spain. She is the former Director of the Administrative Law Department at UNED. She teaches administrative law and urbanism, and her research (9 books and more than 30 articles) is mainly focused on comparative and historical perspectives of property rights and planning law.

Demetrio Muñoz Gielen is a part-time lecturer at the IHS Erasmus Rotterdam University and the Radboud University Nijmegen, in the Netherlands. There he teaches land policies and urban land economics. He is also a part-time public officer at the Municipality of Purmerend, where he designs and implements strategies on developer obligations and public-private partnerships.

Claire Poitras is Professor of Urban Studies at the Centre Urbanisation Culture Société of the Institut National de la Recherche Scientifique in Montreal, Canada and the Scientific Director of the Villes Régions Monde inter-university network. Her areas of research include urban and metropolitan history, the built environment and urban technical networks.

Bill Randolph is a professor and Director of the City Futures Research Centre, Faculty of Built Environment at UNSW Sydney. He leads a research team specialising in housing policy, housing markets and affordability, urban renewal, city data and analytics, urban well-being and metropolitan planning policy issues.

Glen Searle is Honorary Associate Professor of Planning at the Universities of Sydney and Queensland. He was formerly Director of the Planning Programs at the University of Technology Sydney and the University of Queensland. He has held urban policy and planning positions at the UK Department of the Environment and at the New South Wales departments of Decentralisation and Development, Treasury, and Planning, where he was Manager of Policy.

John Sheehan was described by the American Bar Association in 2010 as the foremost Australian authority on compensation assessment arising from compulsory acquisition of native title. He was one of the 35 members of the Expert Meeting convened in 2010 by the UN Food and Agriculture Organisation to report on *Land Tenure Issues and Requirements for Implementing Climate Change Mitigation Policies in the Forestry and Agriculture Sectors*. In 2013, he chaired the Australian Property Institute's response to the NSW Government's Just Terms Compensation proposals. He is a former Acting Commissioner with the NSW Land and Environment Court, and previously a Member of the Aboriginal and Torres Strait Islander Land Tribunal of Queensland.

Tejo Spit is Full Professor in Urban and Regional Planning at Utrecht University, but also holds a position in the Land Use Planning Group of Wageningen University & Research, the Netherlands. He specialises in land policy, planning methodology, infrastructure planning and administrative aspects of spatial planning.

Ed Sullivan teaches Planning Law at Portland State University, Lewis and Clark Law School and Willamette University College of Law, USA. He is a member of the American Planning Association's Legislative and Policy and Amicus Curiae Committees, co-authors a monthly column on planning law for the Portland *Daily Journal of Commerce*, and is the author of more than 60 law review articles relating to constitutional, administrative and planning law. Ed is a retired partner in a Pacific Northwest law firm and is a former chair of the American Bar Association's Section of State and Local Government Law.

Laurence Troy is a research fellow and lecturer at the City Futures Research Centre, Faculty of Built Environment at UNSW, Sydney. His research focuses on housing markets, urban governance, urban renewal and socio-economic outcomes in Australian cities.

Yunpeng Zhang is a postdoctoral research fellow at the Division of Geography and Tourism, KU Leuven, Belgium. His work was previously funded by the European Research Council (Grant Number: 313376) and is currently funded by the Research Foundation – Flanders (FWO).

Foreword

Real Property Rights are central to the economy and provide a legal framework for how society (be it developed or customary) relates to land and buildings. Property rights are both institutional arrangements and social relations. We need to better understand property rights to ensure sustainable societies, careful use of limited resources and sound ecological stewardship of our land and water.

Land conflict is all around us – from corporate and political corruption over land dealings in the developed world, to land grab in developing countries, to compromised indigenous property rights, to resource exploitation. At a time where global food security, water security and shelter are paramount, an understanding of property rights is key to sustainability.

Contemporary property rights theory is dynamic and this series strives to engage thinkers who are prepared to step beyond their disciplinary limitations. 'Property Rights' is a broad term that is fundamentally about social relations. Real property rights, obligations and restrictions can be found in and change across the full range of human societies, both in time and space. Property rights research has emerged from a broad range of disciplines including (but not limited to) archaeology, anthropology, ethics, sociology, psychology, law, geography, history, philosophy, economics, planning and business studies. What makes this series special is that it facilitates a transdisciplinary approach to understanding property rights and specifically promotes heterodox thinking.

In *Compulsory Property Acquisition for Urban Densification* Glen Searle, as editor, has brought together a group of leading thinkers on the contemporary city to address nine carefully crafted questions on how to manage the property rights, obligations and restrictions impacted by densification. 'Urban densification' has entered the property rights lexicon as an urban planning concept that implies the most socially and environmentally sustainable way to house ever increasing urban populations is by providing homes that are more compact and closer together whilst preserving public 'common' spaces and notions of community. However, the social and environmental aspirations of sustainability cannot be taken in isolation from the economic reality that faces local, state and national governments in managing urban growth. It is also these multiple levels of government that hold statutory, constitutional and state political power to engineer individual private property rights under the guise of the wider public interest and long-term political 'good'.

At worst, visions of political good and public interest may only be as long term as the current election cycle. At best, they achieve a longer-term vision for the comprehensive planning of cities to prepare for and adapt to significantly increasing populations occupying the same geographical space in the short to medium term. This dynamic evolution of our urban areas necessitates a shift in both the public-private relationships as well as the spatial relationships of the city. As human constructs, property rights relationships need to be equally dynamic and flexible in adapting to growing cities and evolving approaches to occupying and sharing the urban landscape sustainably. Our individual real property rights are only as strong as the institutions (of government) that secure and guarantee them, and this book explores the social, environmental, economic and political tensions that are central to managing urban densification in the public interest.

The reality, and ramifications, of the densification aspiration in urban planning are explored in this edited volume through international examples of the property rights *tensions* involved in attempting to foster denser, more sustainable cities through the compulsory acquisition (eminent domain) of property in target areas. The case studies from Europe, North America, eastern Asia and Australia show how well, or not, property rights have been recognised in each country. The case studies explore the significance of local legal frameworks and institutions in accommodating property rights in the densification process.

The need for densification has arisen in order to reduce the infrastructure, transport and land costs for urban growth so as to make the expansion of cities as sustainable as economics and politics permit. While the benefits of densification might seem clear in most cases, the use of compulsory acquisition to gain these benefits is a contested and potentially fraught process. Governments have been willing to compromise societal property rights via compulsory acquisition in order to minimise expenditure. A central theme of the book explores the degree to which compromised real property rights involved in the densification process are justified in terms of the wider public infrastructure and other costs that are intended to be saved. The underlying public interest issue explored by the book examines the equity of compromising individual property rights for the broader goal of achieving more sustainable cities via densification.

The book is unique in terms of setting aspects of current just city discourse against sustainability discourse, providing a valuable resource for readers with an interest in environmental law, planning and geography, urban and housing policy makers in national and sub-national governments and think tanks, and lawyers involved in cross-national compulsory property acquisition cases.

The comparative case studies provide extensive international analysis of the degree to which sustainable city policy in the form of densification is traded off against property rights justice. The book significantly expands the literature on compulsory property acquisition in the public interest beyond traditional public interest concerns such as provision of land for major transport routes, to a major new public interest concern, achieving sustainable cities via densification. Through the worthy contributing chapters in *Compulsory Property Acquisition for Urban Densification*, Glen Searle synthesises his analytical framework to offer a broader international understanding of the role of property rights amid the milieu of competing social, environmental, economic and political pressures in the rapidly expanding contemporary urban landscape. Enjoy.

Spike Boydell, General Editor
Sydney, April 2018

1 Introduction

Glen Searle

Densification has emerged as a widely applied key policy for achieving sustainable urbanisation. The benefits can justify government intervention to enforce more rapid densification. This book surveys one such intervention to generate densification: compulsory property acquisition (also known as expropriation, or embodied under eminent domain, condemnation or physical taking in the US). The book is intended to compare and contrast the range of compulsory property acquisition laws and practices across the globe as applied to the goal of increasing densification. It thus differs from Alterman's (2010) survey of cross-national law and practice in compulsory property acquisition in that it focuses on the ways in which such a policy is specifically used to facilitate urban densification.

Compulsory property acquisition is a politically contentious and controversial policy, as it involves some extra-market taking of property rights. It has generated significant protests in the US, in particular (Kohn, 2004; Staeheli and Mitchell, 2008). But in some cases, notably where there are serious market impediments such as highly fragmented property ownership, it may be the only real way in which socially desirable densification can be made to happen. Thus the conditions under which this is effected in order to achieve densification are of significant interest to all those who are involved in, and who analyse, urban policy-making. The book is intended to provide a cross-national perspective on the circumstances and manner under which compulsory acquisition for densification takes place. The fundamental issue explored by the book is: to what extent is property rights justice compromised to achieve more sustainable cities via densification? The case studies from Europe, North America, eastern Asia and Australia show how well, or not, property rights have been recognised in each country in relation to compulsory acquisition for densification, and the contexts that have produced those outcomes.

Densification, as a central element of smart city and compact city policies, is being sought by governments around the world because of the perceived public benefits it brings. Foremost among these are savings in infrastructure costs, with shorter pipes, wires and roads needed to accommodate a given increase in population. Another is savings in energy costs and greenhouse gas emissions with shorter trips and increased walking, cycling and public transport use (Newman and Kenworthy, 1989), though extra energy consumption in high rise buildings can partially offset transport energy savings (Rickwood *et al.*, 2008). A third public benefit of densification comes from savings in the consumption of land for urban development, thus reducing per dwelling housing costs and preserving natural vegetation and agricultural production (the latter being the major driver of China's Link Policy, described in Chapter 9 in this book). In addition, the replacement of older housing by new housing through densification can mean that living standards are significantly improved, as shown in the chapters on China and Singapore in this book.

Many, if not most, of these benefits do not have a market price that reflects their worth. In economic terms this means they are positive externalities (that is, they are external to being priced by the market). Property rights should be rearranged to produce these benefits as long as the benefits produced exceed the costs of doing so (this is an implication of the 'corollary of the Coase theorem' – see Coase, 1960; Lai and Lorne, 2015). In theory, property rights could be controlled by regulation so that these benefits are produced by normal property development. However the transaction costs of, first, precisely and completely identifying the nature of externalities arising from every development and, second, enacting and policing regulations that 'internalise' these externalities within each development, are highly likely to be excessive. Planning regulations commonly attempt to control major externalities on a one size fits all basis, but without much knowledge of how the value of positive externalities generated (or negative externalities saved) by regulations compares with the extra costs to developers. Compulsory acquisition offers the possibility of simpler and lower transaction costs for controlling property rights to achieve densification.

The distinction between compulsory acquisition and planning regulations to produce densification can also be made at a more conceptual level. In particular, the use of compulsory acquisition can be viewed as converting, in the first instance, private property rights into a publicly owned commons. This commons is then converted back, in the process of densification, into new property rights and also into new commons (new road layouts, etc. but also broader commons such as clean(er) air produced by savings in transport usage and conservation of natural vegetation or farmland).

The case for compulsory acquisition to achieve the public benefits of densification rests to a large extent, following the logic of the corollary of the Coase theorem, on the argument that the costs of achieving densification by this process will be less than the costs of doing so via development regulations and controls. One critical advantage of compulsory acquisition is that it can achieve economies of scale and timing in generating public benefits. Unless the precinct to be densified has just one owner of the properties within it, the impact of regulations and controls will depend on when each property owner wishes to (re)develop their property. This means that the provision of new infrastructure for the increased population will be delayed, or provided in piecemeal fashion, according to where and how many properties are made available for densification. Such advantages need to be weighed against the costs of compulsory acquisition. Apart from the money costs of acquisition (which in nearly all cases will be more than offset by the increased value of property after densification) and the issue of whether a fair price is paid to property owners, there is a range of less tangible costs such as administrative and judicial costs (including the costs of due process and justice, or the intangible costs if these are lacking), and social and psychological costs involved from households being unwillingly uprooted from their place of residence and from an existing community. A major goal of this book is to identify the various ways in which such costs of compulsory acquisition are identified and accounted for in the process of densification.

Overall, then, the case for compulsory acquisition for densification being in the public interest rests on the claim that the public benefits thus obtained outweigh the costs involved. Nevertheless, the expropriation of standing property rights means that such a policy has political risks from the opposition of those losing their property. The extent to which public benefits are outweighed by economic, social and psychic costs to the property owners concerned will depend to a great degree on the fairness and justness of compensation given for expropriation. The range of compensation rights accorded property owners in different countries has been surveyed by Alterman (2010). There is considerable variation in the extent to which different aspects of property rights (such as the nine distinct rights identified by Honoré (1961) as typically present in full ownership of land) are properly compensated. This particular issue is also a focus of this book, with the chapters on Spain and China, for example, illustrating how densification can allow compensation via the allocation of new property rights, albeit that such compensation might not always fully compensate for lost property rights. The very nature of property rights that should be compensated also shows significant cultural variation. Thus aspects involving place attachment and liberty, for example, have different emphases between UK and Australian tradition on the one hand, and US tradition on the other (Bartel and Graham, 2016).

The issue of property rights compensation is a starting point for wider consideration of critical questions concerning the ways in which compulsory acquisition is used to increase urban densification. This leads, first, to the justification in practice and theory for such a process. Related to this, issues arise of the particular kinds of public benefit that are sought by governments from this process. Benefits that are simply encapsulated under a broad 'public interest' criterion, for example, risk acquisitions that do not produce benefits justifying the costs involved, unless supported by procedures that require the public interest to be spelt out in each case. A central set of issues relates to the nature of compensation for expropriated property, such as whether there is full or only partial compensation and the basis for assessing compensation value; whether there is an obligation to provide alternative housing in instances where residential properties are acquired; and whether there is provision for negotiation or appeal rights. Identification of the legal, administrative and development industry contexts of such issues is also important to help understanding of why better or worse examples (particularly in terms of fairness and justice) of compensation come into being. Finally, any limitations on the scope of compulsory acquisition in terms of types or location of densification, and the alternatives for achieving densification generally, can be important for understanding the relative importance of such acquisition in a particular national context.

These issues lead to the following questions that this book aims to explore:

- Under what circumstances is the taking of property rights via compulsory acquisition to increase densification justified in practice and in theory?
- What specific public benefits are put forward by public agencies for compulsory acquisition?
- What role does the development industry play in facilitating, encouraging or promoting compulsory acquisition?
- What compensation or offsets are offered for such property acquisition, and how are they funded/provided? What is the value basis for compensation? Are some land parcels resumed without compensation?
- Does compulsory acquisition require special legislation or planning controls, or is it constitutionally allowed? What appeal rights are allowed? What obligation is there to negotiate?
- Is there a local or national history of compulsory property acquisition by government for a range of purposes?
- Is compulsory acquisition restricted to certain types or locations of densification?
- Where existing housing is acquired, is there any obligation to provide alternative housing in lieu?
- Are other methods also used to directly bring about densification? If so, when is each method (compulsory acquisition and other method(s)) used, and how are choices between them made?

The answers to these questions will depend to a significant degree on the prevailing political ideology in each country. In particular, they will be closely related to the protection that local laws give to property rights. The extent to which such rights are protected will be central to the resolution of compulsory acquisition issues such as appeal rights, amount of compensation and obligation to negotiate. However, within jurisdictions with similar ideology and levels of protection of property rights there can be differing outcomes depending on the specific way in which rights are protected and/or compensated. For example, ideological opposition to compulsory acquisition as reflected in property rights statutes will translate into varying levels of de facto opposition depending on the actual compensation offered, which may not have detailed statutory direction.

To provide a suitable cross-section of international examples for answering these questions, setting the ideological context of property rights in each case, and commenting on the extent to which compulsory acquisition is resisted because of inadequate compensation and/or other reasons, a number of academics in a range of countries with varied legal, social, economic and political contexts were approached to write a chapter reviewing policies at a national level and/or providing case studies. Each contributor to the book was asked to address the above questions in their chapters to the extent that each case study allowed.

Structure of the book

The various international contributions have been grouped into three sections according to the nature and process of the compulsory acquisition being analysed. The first section contains analyses of compulsory acquisition in countries where the public interest rationale is well established (though often contested, particularly in the US), and compensation is usually on just terms (this is more uncertain in Canada), even though actual constitutional protection of property rights may be weak or non-existent (as in the UK and Canada – see Alterman, 2010, 28–30). Chapter 2, by Holtslag-Broekhof, Hartmann and Spit, reviews compulsory land acquisition in the Netherlands, where its use for densification has declined, and is changing from an active role to a more general one of facilitating land use policy. The authors' reflections on the decline in compulsory land acquisition in the Netherlands embody issues relating to effectiveness, efficiency, democratic legitimacy and inherent justice. In the following chapter, Sullivan covers the large and varied terrain of compulsory purchase (eminent domain) for densification in the United States. The extent to which this is used varies widely according to state constitutions and statutes. The necessity for the acquisition to be for a public purpose can be liberally interpreted: local governments may partner with a developer to acquire property for residential construction, sometimes with the developer providing the funds for acquisition.

In Chapter 4, Cox examines current United Kingdom law and practice in compulsory purchase in the context of large scale housing-led urban regeneration projects. Several recent case studies are compared to gain an appreciation of the challenges involved in undertaking a successful compulsory acquisition. Cox concludes that the legal process of compulsory purchase, while being a key tool for achieving regenerations of old public housing estates, is highly complex and in need of reform. In Chapter 5 Albrecht describes German land expropriation for land readjustment as a traditional method for densification. Albrecht argues that the strict constitutional requirements for its use, entailing much effort and considerable legal risks, mean that the instrument is rarely used to bring about densification. A Canadian case study, in inner Montréal, is described in Chapter 6 by Darchen and Poitras. The chapter describes the role of compulsory property acquisition as a central factor in enabling Montréal City to accommodate densification of a former industrial area while also preserving the area's identity and achieving social sustainability.

The next group of chapters analyse arrangements in which acquisition has been accompanied by the allocation of new property rights, including resettlement. As outlined in Chapter 7 by Muñoz Gielen and Lora-Tamayo Vallvé in the case of Spain, landowners whose property is compulsorily acquired are most commonly given title to new properties created through the densification project. A legally binding land use plan provides sufficient statutory basis for compulsory acquisition.

The following chapters look at Chinese and Singapore experience of land dispossession and concomitant higher density resettlement. In Chapter 8, Yungpen Zhang describes the dispossession of thousands of households from sites in Shanghai that have been redeveloped at higher densities. While there has been a general process of compulsory acquisition of inner city dwellings for redevelopment and city-sponsored resettlement of displaced residents to outer suburbs, the treatment of residents thus affected has been problematic. Zhang describes how different classes of residents were subjected to variable compensation and resettlement outcomes, from complete exclusion of migrant workers not registered as urban, to tenants and property owners, where compensation was often inadequate. The following chapter, by Long Cheng, analyses the national Link Policy for the preservation of China's agricultural land supply, under which rural dwelling land is converted into farmland in return for the development of fringe urban lands to accommodate urban growth. The villagers are re-accommodated in new urban dwellings. The chapter investigates the extent to which those being resettled are satisfied with central aspects of the relocation process and its outcomes.

Singapore's system of compulsory land acquisition for intensified urban development is discussed in Chapter 10 by Alice Christudason. She draws attention to features of the system that have allowed the acquisition and redevelopment of extensive areas of private property to meet a housing backlog. These features include no safeguard of property rights in the constitution, the sweeping interpretation of public benefit, and a betterment levy paid by developers when the government on-sells land that it has up-zoned after acquisition.

The next two chapters analyse Australian examples where expropriation for denser development involves unjust or *ex ante* illegal processes. In Chapter 11, Laura Crommelin and colleagues analyse property rights issues associated with a New South Wales law that allows termination of apartment leases with the approval of 75 per cent of owners in the apartment block to allow denser redevelopment, forcing the remainder to sell. They argue that this can be considered as de facto compulsory acquisition by private interests. In the following chapter, Sheehan and Glinka discuss local compulsory property acquisition law in the context of a Sydney case in which a municipal council compulsorily acquired land and transferred it to developers as part of the creation of a high density regional focus. A landowner appeal on the basis of non-consensual resale and lack of 'public purpose' was successful, leading to a council-instigated parliamentary amendment to change the land acquisition act to authorise the acquisition.

The concluding chapter reviews the ways in which the international case studies in the preceding chapters bring about densification, especially in regard to the fairness of the processes used and outcomes achieved. In this, the central questions the book poses about compulsory acquisition for densification are used to draw out the main conclusions from the case studies. The lessons of the case studies highlight advantages and disadvantages of different national laws and practices (c.f. Alterman, 2010) in using compulsory acquisition to bring about densification.

References

Alterman R (2010) *Takings International: A Comparative Perspective on Land Use Regulations and Compensation Rights*. Chicago: American Bar Association Publications.

Bartel R and Graham N (2016) Property and place attachment, *Geographical Research*, 54(3), 267–284.

Coase R H (1960) The problem of social cost, *Journal of Law and Economics*, 3, October, 1–44.

Honoré T (1961) Ownership. In: Guest A G (ed.) *Oxford Essays in Jurisprudence*, 107–147. Oxford: Oxford University Press.

Kohn M (2004) *Brave New Neighborhoods: The Privatization of Public Space*. New York: Routledge.

Lai L W C and Lorne F T (2015) The fourth Coase theorem: State planning roles and spontaneity in action, *Planning Theory*, 14(1), 44–69.

Newman P and Kenworthy J (1989) *Cities and Automobile Dependence: An International Sourcebook*. Aldershot, England: Gower.

Rickwood P, Glazebrook G and Searle G (2008) Urban structure and energy: A review, *Urban Policy and Research*, 26(1), 57–81.

Staeheli L and Mitchell D (2008) *The People's Property: Power, Politics and the Public*. New York: Routledge.

2 Compulsory acquisition in the Netherlands

Sanne Holtslag-Broekhof, Thomas Hartmann and Tejo Spit

Introduction

One can assume at first glance that compulsory acquisition for urban densification in the Netherlands works similarly to other countries elaborated in this volume. However, as Alterman puts it, the "minor nuances" of instruments can make major differences in their application (Alterman, 2016). This contribution elaborates the "minor nuances" along rather specific aspects of compulsory acquisition in the Netherlands. They are presented along elements of public policy analysis (Salamon, 2000): legitimacy, procedural efficiency and effectiveness, and justice of the instrument in the Netherlands.

Legitimacy is the way the public intervention is justified. Legitimacy can be derived from the institution itself (input legitimacy), via the process (throughput legitimacy), or by the result of an intervention (output legitimacy) (Schmidt, 2013). Procedural efficiency and effectiveness depend on the effort (i.e. time and money) invested in achieving a certain goal (Needham, 2007). Effectiveness relates to the degree a certain instrument contributes to achieving the planning goal in question. This includes a brief outline on the Dutch context of urban densification. Justice, then, is about the distribution of costs and benefits. For compulsory acquisition, this especially affects the compensation for the expropriation to the previous landowner.

These four criteria (see also Hartmann and Spit, 2015) provide an analytical framework and overall structure for the assessment of Dutch compulsory acquisition for urban densification. In addition, the context in which an instrument operates matters substantially (Gerber *et al.*, 2018). Before diving into this analysis, it is essential to understand the wider context of Dutch active land policy to grasp the way the instrument of compulsory acquisition functions in the Netherlands. This helps to contextualize the effectiveness of the instrument in the canon of Dutch instruments of land policy.

Active land policy in the Netherlands

Dutch land policy is among the most extreme forms of active land policy. This means that municipalities actively and strategically acquire land, designate it as building land and – often in collaboration with real-estate developers – produce housing (Buitelaar, 2010). These houses are then sold to developers, businesses or individuals (Louw *et al.*, 2003; Needham, 2014). In the Netherlands, planning agencies take initiative to implement plans, rather than wait passively for market initiatives to occur (Needham, 2007). This explains why the efficiency and effectiveness of Dutch land use planning has been admired (Dieterich, 1998) and even referred to as a planner's paradise, for a long time (Faludi and van der Valk, 1994).

This means that Dutch municipalities act in similar ways to real-estate developers – taking risks and profits by investing in land – as well as behave like public authorities, i.e. granting building permits and land use plans. Consequently, "the majority of land-use plans show more similarity with a contract in which informal agreements are formalized" (Buitelaar, 2012, 215). This behaviour is not only observed in the Netherlands, but is also custom in several other countries such as Finland and Switzerland (Lefcoe, 1978). How has such a land policy emerged in the Netherlands and what are the implications for compulsory acquisition?

Dutch land policy developed out of a substantial shortage of building land after the Second World War. This shortage was the impetus for active land policy. Municipalities needed to produce a massive amount of housing – especially social housing – in a relatively short period of time. So, a very efficient and effective land policy was needed. Active land policy provided a solution. As social housing was generously subsidized by the government, the real estate market became less interesting for private developers. Over time, this led to a quasi-monopoly of the municipalities on the housing market. Later, this quasi-monopoly was weakened due to budget restrictions and other constraints. This brought real estate developers back in the game, who collaborated with municipalities in the development of building land in the past few decades. In practice, active land policy developed as the standard way of developing land in the Netherlands (Needham, 1997; Louw *et al.*, 2003; Hartmann and Spit, 2015), whereas in recent years various forms and models of active land policy with different roles and responsibilities for private businesses and individuals emerged (e.g. collective trust models, concession models, and so on).

The recent economic crisis from 2007/2008 actually started as a real estate crisis in the USA. One effect was the collapse of the private housing market. This had fundamental consequences for Dutch municipalities, which had invested enormous amounts of public money in anticipation of climbing real estate markets before the crisis. Small and middle-sized municipalities in the suburban agglomerations of large cities suffered major losses due to the crisis. This situation led to huge disturbances of active land policy and reconsiderations. In this context, existing instruments of land policy were questioned and discussed. A pending major reform in the years after the crisis replaced the existing planning law (*Wet ruimtelijke ordening*) with a comprehensive Law on the Environment (*Omgevingswet*, which introduces new instruments and is expected to fundamentally revise Dutch land policy). However, a current trend can be observed showing that active land policy is once again being applied on large scales. But the crisis led to an active debate on instruments of land policy. For these reasons, a lively debate exists about land policy (van der Krabben and Needham, 2008; Buitelaar, 2010; van Straalen *et al.*, 2014; Woestenburg *et al.*, 2014; Hartmann and Spit, 2015; Buitelaar and Bregman, 2016; Holtslag-Broekhof, 2016).

This background is important to understand the mechanism and the strategic use of compulsory acquisition in the Netherlands, especially in urban densification processes. Active land policy involves the use of public power and instruments to acquire land for the designated land use (Needham, 1997; van der Krabben and Needham, 2008; Buitelaar, 2010).

Urban densification policies in the Netherlands

In the Netherlands, urban densification became a formal policy in 1985, after it was accepted by the Dutch parliament in an official national memorandum on spatial planning (Ministerie van VROM, 1985). Earlier, the municipality of Amsterdam advocated densification policies and put them into practice even before they became official national policy. In fact, it was the result of an intensive discussion on the effects of the earlier policy on urbanization, which was called 'bundled de-concentration'. As an effect, larger cities started losing their population. More specifically, the well-to-do households left the larger municipalities in the Randstad in vast numbers, and cities began to deteriorate rapidly. Official stimulation of urban renewal, involving all tiers of government in combined efforts, tried to turn this downward spiralling effect in the opposite direction (Spit and Zoete 2016, 42–43). Urban densification policies became more urgent, as the capacity needed to sustain shops and facilities in cities increased enormously. To keep those thriving within the urban areas, an increase of the (wealthy) population became an urgent matter. Awareness also started to grow about public transportation as an alternative to car use, so densified urban areas proved to to offer a prime role in this (Richardson and Bae, 2004).

Since then, densification policies have never left the political agenda as a major spatial strategy. This is reflected in national policy documents, as well as in local spatial planning documents (Spit and Zoete, 2016, 42–51). The result of prioritizing densification policies for more than 30 years, however, was that in the urban areas, the locations that were relatively easy to (re)develop were becoming fewer and fewer. Now, the only locations left in urban areas to (re)develop all have complications. This has major consequences for the effectiveness and efficiency of these policies. They are visible all over the country. In terms of quantity, one wouldn't expect that so many dwellings could be built within the urban areas, not only because of the complexity, but because there were not enough (large) suitable areas to (re)develop. Second, the development of these locations has become increasingly expensive. So either the profits of (re)development must be higher or subsidies must cover them. Third, it takes more and more time to (re)develop those areas since they are more complex, which then has consequences for the quantity available. Fourth, the (re)development of complex locations will demand more expertise on specific issues related to the location itself (e.g. soil pollution, or road attainability, etc.), but also with the process. The latter refers not only to the management of a complex redevelopment process in which several types of expertise must be combined, but also to the issue of fragmented ownership of the land involved. Fragmented ownership of property is a regularly occurring phenomenon in urban areas and draws heavily on specific instruments (e.g. compulsory acquisition) to organize the (re)development process (see Woestenburg *et al.*, 2018).

In summary, the densification process of urban areas in the Netherlands has grown into very complicated development processes, which draw heavily not only on content related expertise, but also on process organizing skills. As a result, the use of instruments such as compulsory acquisition has become an increasingly central part of each development project or process.

Compulsory acquisition in the Netherlands

The current Dutch legislation for compulsory acquisition[1] originated in 1852 and has only been slightly adapted since. Compulsory acquisition is perceived as the most far reaching legal instrument of land policy (Overwater, 1999). The procedure of compulsory acquisition can roughly be divided in an administrative and a juridical procedure. During the administrative phase, the ground for and exact location of expropriation are determined. In the juridical procedure, the expropriation is formalized and the compensation amount is determined (Overwater, 2002; Kalt, 2006). Compulsory acquisition is only justified under a number of strict circumstances. First, a legal compulsory acquisition procedure must be preceded by sufficient voluntary negotiations on the purchase of the land. Second, compulsory acquisition can only be used to pursue the public interest. Third, all affected parties should be financially compensated for all financial harm that is caused by the compulsory acquisition. In practice, the majority of compulsory acquisition cases end with a 'voluntary' purchase, before the juridical procedure has even started. Therefore, the ability of compulsory acquisition to put pressure on voluntary negotiations is a major strength of the legislation.

Analysing Dutch compulsory acquisition

Dutch compulsory acquisition is distinguished in three aspects from many other countries. It is rooted in Dutch active land policy and densification policy. These particularities may be further explained in the following paragraphs.

Legitimacy – only in 'public' interest

Compulsory acquisition can only be used to pursue the public interest, which is reflected in the first article of the compulsory acquisition act (article 1.1). In the clarification note ('*memorie van toelichting*') of the law, Thorbecke, the founding father of the Dutch law system, states that the meaning of public interest should be determined by the legislator (Sluysmans, 2011).

The definition of what is in the public interest is crucial, however. Whereas in other countries (such as Germany) a rather narrow definition of the public interest is applied, the Dutch interpretation of public interest is rather broad. Under the current legislation, purposes for compulsory acquisition are registered in nine 'principles for compulsory acquisition'. For spatial planning purposes, the principle of 'infrastructure' (2a) and the principle of 'spatial development and housing' (4) are generally used. The latter can be used as ground for compulsory acquisition for land use plans, as long as this sufficiently serves the interest of public housing, the interest of spatial developments, or the public interest (Overwater, 2002). This results in the ability to use compulsory acquisition for each land use, as long as it can serve one or more of the above-mentioned interests and is registered in a land use plan. This is not limited to spatial aspects, but may also be to the economic profit of the municipality. Dutch active land policy allows granting the implementation of the land use plan to a private party. The fact that this private party may profit off the development does not imply that the development cannot simultaneously be in the public interest. Landowners can even be expropriated for private land use with an indirect public interest.

For example, in 2010, the owner of a pancake restaurant in the small town of Leiderdorp was expropriated for an IKEA warehouse. The municipality had a cooperation agreement with the warehouse, which included the obligation of the municipality to acquire extra land, via compulsory acquisition, if necessary. As such, the municipality expropriated the functioning pancake restaurant for the IKEA warehouse. There was commotion over this case, as the compulsory acquisition was argued to not be in the public interest, but rather in the interest of the IKEA warehouse. In research by the Dutch national ministry, the researchers concluded that the compulsory acquisition was first executed to effectuate the public-private agreement, and not to realize the zoning plan. However, the compulsory acquisition was approved by the court[2] based on the land use plan of the municipality that included the development of retail in the area. The court argued that since the municipality of Leiderdorp already had a zoning plan to develop the area before they came to an agreement with IKEA, the compulsory acquisition was not only in the interest of the IKEA, but also had a public interest. Moreover, the Crown argued that the establishment of the IKEA warehouse was also dictated by the municipal financial-economic policy which services the public interest of the municipality (De Roos, 2011). The fact that the public interest was possibly not the first interest or argument to expropriate the private land owner was not relevant for the decision of the court.

The legitimacy of compulsory acquisition in the Netherlands is thus legally grounded in the precondition of a public interest, while in practice the interpretation of the public interest is broad and becomes increasingly broader.

Efficiency and effectiveness – procedure ensures quick realization of goals

The legal compulsory acquisition procedure must always be preceded by sufficient voluntary negotiations on the purchase of the land. If voluntary negotiations would not result in the purchase of the land, the government summons the land owner with a final compensation offer. If the landowner does not accept this offer, the compulsory acquisition is decided in court. The procedure in court can be lengthy, as both the government and the land owner are allowed to use cassation appeal (a legal procedure in which one can appeal at the highest judging institute against the verdict of a lower judge). In 1972, a revised procedure for compulsory acquisition was introduced, ensuring a significant procedural acceleration. Using this procedure, public authorities could start realizing the plans after the expropriation had been determined in court, but before the court had ruled over the exact amount of the compensation (Sluysmans, 2011). This increased the efficiency of the compulsory acquisition procedure. Today, this accelerated procedure is almost always applied. The procedure enables public bodies to become the legal owner of the land at the start of the juridical procedure. From there, the juridical procedure for the exact amount of the full compensation could continue for years, while plan realizations had already started. However, the costs of a court procedure of compulsory acquisition remain significant. Process costs could be part of the compensation (see next paragraph) if the costs were reasonable and remained within reasonable proportion (Scheltema *et al.*, 2017). Because the legal process costs can be relatively high, especially if the value of the expropriated land is low, it may be much more efficient to acquire the land for a higher price, to prevent a court process of compulsory acquisition. However, the split procedure in the expropriation and the compensation aspects illustrates the pragmatic approach to compulsory acquisition in the Netherlands.

Justice – full compensation

The basis of the compulsory acquisition compensation regulations is that all affected parties should be fully indemnified. More specifically, they should be financially compensated for all financial harm that affected parties suffer 'directly and necessarily' from the loss of their property caused by the compulsory acquisition (De Roos, 2010). The compensation is limited to a financial compensation. Although the government is obliged to offer alternative property during the voluntary negotiations, the owner has no right to alternative property or housing. If the voluntary negotiations fail and the expropriation is proclaimed in court, the owner only receives a financial compensation for the loss of the property.

A full financial compensation for all harm that the owner suffers directly and necessarily from the loss of his property (article 40) includes:

- market value of property;
- decreased value of the remaining property;

- loss of income due to compulsory purchase;
- other financial damage.

The market value of the property is defined in Dutch legislation as the price of a presumed sale in the free commercial market between a reasonably acting buyer and seller (article 40b of the Dutch compulsory acquisition act). This definition implies that hope value – based on the future use of the land – is included in the market value, if the hope value was already present when compulsory acquisition was pronounced. This differs from most other countries in which only the existing use of the land is compensated (Gerber *et al.*, 2018). However, the appraisal of hope value is not easy, especially if comparable reference transactions are amiss.

Objective appraisers estimate the amount of compensation on behalf of the government or the legal owner. This is a complex practice that requires a lot of experience and appraisers need to be legal experts as well to determine the full compensation (Holtslag-Broekhof, 2016).

Conclusion

Dutch compulsory acquisition is only justified under a number of strict circumstances: there have been sufficient voluntary negotiations, the acquisition is in the public interest, and the owner receives sufficient financial compensation. These preconditions are recorded in Dutch compulsory acquisition law. When it comes to the number of actual court cases, the use of compulsory acquisition in the Netherlands is quite limited. However, the value of compulsory acquisition to prevent these court cases, by pushing land owners to sell voluntarily, is more important than the value that can be attributed to the limited number of court cases. As such, the instrument and its principles have – despite the limited number of court cases – a strong influence upon public-private land transactions.

The choice for the implementation of compulsory acquisition is a deliberation between efficiency and purposefulness on the one hand, and democratic legitimacy and the right to private property on the other hand. For Dutch urban densification tasks, the choice to implement compulsory acquisition must be made by the municipality. This is a political choice that the city council makes. The number of enforced compulsory acquisitions in practice is limited, not only because the procedure is complex and can be costly, but also because most politicians perceive compulsory acquisition as an *ultimo ratio*; something that you should only use if nothing else suffices. Although every owner has a right to full indemnification of the property, there is no right to replace land. Moreover, the combination of full indemnification of all financial harms and the potential costs of a legal procedure make the instrument in some situations costly for municipalities to use. The instrument is therefore only sporadically used, especially in urban areas where politicians have relatively tight relations with their citizens. After years of successful urban renewal in inner Dutch urban areas, most 'easy' locations have been used in the past. Places that are complex to redevelop have remained for future densification tasks. Locations that remain are often small, have high land values, and are often complex in terms of ownership rights, parcel structure or soil pollution.

The choice to enforce compulsory acquisition in practice can only be understood in relation to the alternatives municipalities have to the use of compulsory acquisition. The choice to use compulsory acquisition for urban densification is never a choice on its own, but is always part of a broader political deliberation on the urban planning goals, available resources, time planning and instruments of land policy. Municipalities have various instruments to implement urban plans that could be placed in a hierarchy of impact upon the right to private property. At the bottom of this hierarchy, voluntary land readjustments and voluntary land acquisitions are found. These instruments depend fully on the voluntary participation of the involved land owners. Pre-emption rights (i.e. a formal instrument that entitles the public authority for the first option to buy a property under certain conditions, see Gerber *et al.*, 2018) and land consolidation stand a bit higher on the ladder. At the top of the hierarchy of impact upon private property stands compulsory acquisition. The use of compulsory acquisition is part of active land policy. The ranking is shown below.

1 Compulsory acquisition;
2 Land consolidation;
3 Pre-emption rights;
4 Voluntary land readjustment;
5 Voluntary land acquisition.

However, in practice, the shadow of a possible compulsory acquisition can push land owners toward a voluntary sale of their land. Therewith, all kinds of 'light' and perceived interactions between these instruments (willingly or not) can influence its use. Consequently, this shows the importance of studying these instruments in relation to each other.

In the Netherlands, municipalities increasingly opt for a passive land policy instead of the previously mentioned active land policy, which mostly excludes the use of compulsory acquisition. The active land policy has been a long-term standard practice by Dutch municipalities. After the economic crisis of 2007/2008, this practice has been changed and municipalities search for more facilitative approaches to steer spatial developments, especially in urban areas where land prices are high. However, as mentioned before, compulsory acquisition can also be used as part of a passive land policy to acquire land as part of an agreement with a private developer.

Countries that use active land policy similar to the Netherlands, such as Finland and Switzerland, have similar dilemmas regarding the use of instruments of land policy. Finally, compulsory acquisition in the Netherlands is effective and efficient due to the pragmatic approach to legitimacy. It is, however, barely used in practice due to the high procedural costs but also because it is a very interventionistic instrument. This does not fit the Dutch collaborative planning practice, which builds largely on collaboration – in the vein of the Dutch polder model.

Notes

1 Currently, a major revision in Dutch planning regulation is pending. This revision includes several procedural changes in the law for compulsory acquisition. In this chapter, we will not extensively explain and react to the new law for compulsory acquisition, as the main preconditions for compulsory acquisition remain unimpaired.
2 http://jure.nl/ECLI:NL:RBSGR:2012:BW4121.

References

Alterman R (2016) Planners' beacon, compass and scale: Linking planning theory, implementation analysis and planning law. In: Haselberger B (ed.) *Encounters in Planning Thought: 16 Autobiographical Essays from Key Thinkers in Spatial Planning*, 260–279. London: Routledge.

Buitelaar E (2010) Cracks in the myth: Challenges to land policy in the Netherlands, *Tijdschrift voor Economische en Sociale Geografie*, 101(3), 349–356.

Buitelaar E (2012) The fraught relationship between planning and regulation: Land use plans and the conflicts in dealing with uncertainty. In: Hartmann T and Needham B (eds) *Planning by Law and Property Rights Reconsidered*, 207–218. Farnham: Ashgate.

Buitelaar E and Bregman A (2016) Dutch land development institutions in the face of crisis: Trembling pillars in the planners' paradise, *European Planning Studies*, 24(7), 1281–1294.

De Roos J (2010) Kroniek onteigening 2009-II, *Tijdschrift voor agrarisch recht*, 7–8, 253–259.

De Roos J (2011) Kroniek onteigening 2010, *Tijdschrift voor agrarisch recht*, 71(9), 359–367.

Dieterich H (1998) Bodenpolitik in Deutschland, *Neue Zeitschrift für Miet- und Wohnungsrecht*, 1998(8), 289–295.

Faludi A and van der Valk A (1994) *Rule and Order: Dutch Planning Doctrine in the Twentieth Century*. Dordrecht: Springer Science.

Gerber J-D, Hartmann T and Hengstermann A (2018) *Instruments of Land Policy: Dealing with Scarcity of Land*. Abingdon, Oxon: Routledge.

Hartmann T and Spit T (2015) Dilemmas of involvement in land management – Comparing an active (Dutch) and a passive (German) approach, *Land Use Policy*, 42, 729–737.

Holtslag-Broekhof S (2016) *Dealing with Private Property for Public Purposes: An Interdisciplinary Study of Land Transactions from a Micro-Scale Perspective*. Doctoral dissertation, Wageningen University.

Kalt N (2006) *Grondzaken Doen. Handen en Voeten aan het Grondbeleid*. Alphen aan den Rijn: Kluwer.

Lefcoe G (1978) When governments become land developers, *Urban Law and Policy*, 1, 103–160.

Louw E, van der Krabben E and Priemus H (2003) Spatial development policy: Changing roles for local and regional authorities in the Netherlands, *Land Use Policy*, 20(4), 357–366.

Ministerie van VROM (1985) Structuurschets stedelijke gebieden (deel e). Den Haag: Tweede Kamer der Staten Generaal 1985–1986, 18048, nrs. 48–49.

Needham B (1997) Land policy in the Netherlands, *Tijdschrift voor Economische en Sociale Geografie*, 88(3), 291–296.

Needham B (2007) *Dutch Land Use Planning: Planning and Managing Land Use in the Netherlands, the Principles and the Practice*. Den Haag: Sdu uitgevers.

Needham B (2014) *Dutch Land-Use Planning: The Principles and the Practice*. Aldershot, Surrey: Ashgate.

Overwater P S A (1999) Onteigening en een beroep op zelfrealisatie binnen het kader van een nieuw, sturend gemeentelijk grondbeleid, *Tijdschrift voor agrarisch recht*, 12, 631–644.

Overwater P S A (2002) *Naar een Sturend (Gemeentelijk) Grondbeleid. Wie de Grond Heeft, die Bouwt*. Alphen aan den Rijn: Kluwer.

Richardson H W and Bae C-H C (2004) *Urban Sprawl in Western Europe and the United States*. Aldershot: Ashgate.

Salamon L M (2000) The new governance and the tools of public action: An introduction, *Fordham Urban Law Journal*, 28(5), 1611–1673.

Scheltema M W, Rus-van der Velde M, Procee J S, Baljon L, Haan K E and Bouwman S J M (2017) Kroniek onteigeningsjurisprudentie 2016, *Tijdschrift voor Bouwrecht*, 23(2), 117–126.

Schmidt V A (2013) Democracy and legitimacy in the European Union revisited: Input, output and 'throughput', *Political Studies*, 61(1), 2–22.

Sluysmans J A M A (2011) *De Vitaliteit van het Schadeloosstellingsrecht in Onteigeningszaken*. Doctoral dissertation, Universiteit Leiden.

Spit T and Zoete P (2016) *Planologie. Een Wetenschappelijke Introductie in de Ruimtelijke Ordening in Nederland*. Groningen: Cooperatie in Planning U.A., 42–51.

van der Krabben E and Needham B (2008) Land readjustment for value capturing: A new planning tool for urban redevelopment. *Town Planning Review*, 79(6), 651–672.

van Straalen F M, Janssen-Jansen L B and van den Brink A (2014) Delivering planning objectives through regional-based land-use planning and land policy instruments: An assessment of recent experiences in the Dutch provinces, *Environment and Planning C: Government and Policy*, 32(3), 567–584.

Woestenburg A, van der Krabben E and Spit T (2014) Institutions in rural land transactions, *Journal of European Real Estate Research*, 7(2), 216–238.

Woestenburg A, van der Krabben E and Spit T (2018) Land policy discretion in times of economic turndown: How local authorities adapt to a new reality, *Land Use Policy* (forthcoming).

3 Eminent domain use for densification in the United States

Ed Sullivan

Introduction

Taking a leaf from the movie *Fight Club*, the reader should note that the first rule about eminent domain (as that term is used to describe compulsory purchase in the United States) is that there is no universal rule about eminent domain in the United States. Instead, there is a constitutional baseline that almost always applies to public acquisition of interests in land and many shared commonalities of eminent domain practice and procedure under federal and state law that may not be required.

In the United States, the federal government (and its agencies), the 50 states and the territorial governments (and their agencies, including most local governments) have the authority to use the eminent domain power, which is an attribute of government sovereignty, authorized in state constitutions and detailed in statutory law. Congress and the various state legislatures have frequently delegated the power to exercise eminent domain to their agencies and, in the case of the states and territories, to local governments. Aside from constitutional baselines, the exercise of this power may be undertaken in multiple ways in terms of offers, appraisals, taking immediate possession, the determination of "just compensation", and the like. While these differences are interesting, this chapter will discuss these differences only with respect to the use of the eminent domain power (also called the power to condemn) as it applies to the acquisition of land to increase residential density in a given area.

Constitutional baseline

In its original form in 1787, the United States Constitution was silent as to the power to acquire land. However, the Anti-Federalists who feared an overly strong central government agitated for limits to certain federal powers, proposing instead what is now called the "Bill of Rights" to do so (Levy, 1999). Most of their proposals were adopted as the first ten amendments to the federal constitution, including the Fifth Amendment, which states in relevant part: "No person shall be … deprived of life, liberty, or property, without due process of law; nor shall private property be taken for public use, without just compensation."

The "due process" clause of the Fifth Amendment assures fair procedural treatment of landowners in the eminent domain process and arguably could be used in a procedural context to deal with the fairness of the eminent domain proceedings. That clause may also be used to deal with issues of fairness outside a procedural controversy.[1] However, the application of the due process clause in the federal constitution to the use of eminent domain for densification (and, because many states have included similar language in their own constitutions as well, to state and local eminent domain proceedings) is largely peripheral to the discussion and will not be discussed further.

The "takings clause" of the Fifth Amendment is replicated in one form or another in all state constitutions or statutory law. As originally adopted, that clause was a restraint only against the federal government.[2] After the American Civil War, however, the Constitution was further amended to change the fundamental relationships among citizens and the state and federal governments. In particular the Fourteenth Amendment, especially its "due process clause" provided for certain civil rights to all citizens, including former slaves, to be guaranteed by the federal government. The Fourteenth Amendment provides in material part:

> All persons born or naturalized in the United States, and subject to the jurisdiction thereof, are citizens of the United States and of the State wherein they reside. No State shall make or enforce any law which shall abridge the privileges or immunities of citizens of the United States; *nor shall any State deprive any person of life, liberty, or property, without due process of law*; nor deny to any person within its jurisdiction the equal protection of the laws.
>
> (Emphasis supplied)

For the century following the adoption of the Fourteenth Amendment in 1867, the Supreme Court undertook "incorporating" certain of the restraints placed on the federal governments as also restraining the states (University of Missouri at Kansas City, n.d.). In 1897, the Supreme Court found that the Fifth Amendment Takings Clause was applicable to state and local governments. *Chicago, Burlington and Quincy Railroad v. City of Chicago* (166 U.S. 266 [1897]). Thus, the Takings Clause was a potential issue in all state and local eminent domain cases, although it was an actual issue in very few cases. Moreover, because state and local constitutions and statutes contained similar provisions (sometimes prohibiting the damaging of property, as well as its acquisition unless just compensation were paid), that similarity gave rise to a convergence of views as to how and when eminent domain were to be used.

Public use

Aside from issues of just compensation, the most frequent constitutional objection to the use of the eminent domain power is that of "public use". There is no uniform definition of this term and the many views of what government can or should do as a "public use" under the federal and state constitutions and laws has resulted in a cacophony of judicial opinions that are relevant to the use of eminent domain in providing for densification. Under federal and state law and practice, both public agency determinations of public use and just compensation may be challenged in the appropriate court. As noted below, public use authorization under state constitutional and statutory law is in flux as those authorities change and is reinterpreted.

To begin this discussion, it is important to note that the federal constitution's Taking Clause provides a "floor" of minimum requirements for the use of eminent domain. States may provide additional limitations through their own constitutions or statutory provisions; however, state constitutions and laws may not provide less protection or limitation on eminent domain than the Fifth Amendment.

The relatively recent case of *Kelo v. City of New London* (545 U.S. 469 [2005]) is now the leading case illustrating this distinction and, although binding as to the federal constitution, has caused many states to limit their eminent domain powers voluntarily.

Here are the facts of the case. Susette Kelo was a long-standing resident in an area of the city that was targeted by an urban renewal agency (a governmental corporation tasked with redevelopment) for an employment centre to replace a shuttered military facility. The urban renewal agency developed a plan to attract a potential employer by providing a campus on which that employer would provide work. Kelo's property was at the edge of the proposed campus and there was no particular use described for her property after acquisition. Kelo fought the acquisition through local proceedings, state court proceedings and ultimately in the United States Supreme Court. The case was not a typical eminent domain case, which usually involves money; rather Kelo challenged the very power of the City to acquire her property.

In framing her challenge, Kelo's public interest law firm, the Institute for Justice,[3] focused upon the "public use" portion of the Takings Clause. The case, they said, was not about money; rather it was about the legitimate function of government. While recognizing the power of government to condemn property because a nuisance was present, or because the property was needed for some governmental function (such as a library or a road widening), Kelo argued that economic development was not a "public use". There had been two "public use" cases in the Supreme Court since World War II and both had been decided, without a dissenting vote, in favour of the public agency condemnor.[4]

In the first case, *Berman v. Parker*,[5] Congress had authorized the Washington D.C. government to use urban renewal to redevelop an area of that city that had deteriorated to the point of becoming a health and safety problem. However, Berman owned a store that was not deteriorated and disputed the city's urban renewal authority declaration of blight, which triggered the authorization for the use of eminent domain to acquire all the land in the blighted area (with just compensation) and its resale to a developer who would comply with an urban renewal plan for the area. The court brushed aside Berman's objections based on the conditions of his own property and found the city had the power to acquire property to implement its area-wide urban renewal plan, concluding:

> We do not sit to determine whether a particular housing project is or is not desirable. The concept of the public welfare is broad and inclusive. … The values it represents are spiritual as well as physical, aesthetic as well as monetary. It is within the power of the legislature to determine that the community should be beautiful as well as healthy, spacious as well as clean, well balanced as well as carefully patrolled. In the present case, the Congress and its authorized agencies have made determinations that take into account a wide variety of values. It is not for us to reappraise them. If those who govern the District of Columbia decide that the Nation's Capital should be beautiful as well as sanitary, there is nothing in the Fifth Amendment that stands in the way.

This oft-quoted and sweeping language informed much of the debate over urban renewal for the next half-century.

In the other case, *Hawai'i Housing Authority v. Midkiff*,[6] the Hawai'i legislature attempted to address the imbalance in property ownership due to that state's property history in which the monarchy had owned most of the land and which ownership was succeeded by a land trust. That trust and 72 other individuals and entities owned 47 per cent of the non-government land in the state. The legislature passed the Land Reform Act, allowing the Housing Authority to acquire (with just compensation) fee interests of those lands that were leased for residential use, with the lessees paying the state to acquire ownership rights. This exercise was challenged as not being an exercise of "public use" because it was allegedly for the benefit of the individual lessees who would, in effect, use the state's power of eminent domain to force acquisition of the fee estate. The Supreme Court (with one justice not sitting) unanimously upheld the state law, concluding:

> The State of Hawaii has never denied that the Constitution forbids even a compensated taking of property when executed for no reason other than to confer a private benefit on a particular private party. A purely private taking could not withstand the scrutiny of the public use requirement; it would serve no legitimate purpose of government, and would thus be void. But no purely private taking is involved in these cases. The Hawaii Legislature enacted its Land Reform Act not to benefit a particular class of identifiable individuals, but to attack certain perceived evils of concentrated property ownership in Hawaii – a legitimate public purpose. Use of the condemnation power to achieve this purpose is not irrational. Since we assume for purposes of these appeals that the weighty demand of just compensation has been met, the requirements of the Fifth and Fourteenth Amendments have been satisfied.

Given *Berman* and *Midkiff*, most lawyers would have opined that the City of New London would have no trouble in having its land acquisition affirmed. That was not to be, however, for the ultimate decision that was consistent with *Berman* and *Midkiff* was decided by a 5–4 margin.

Justice Stevens wrote for the majority and approached the case along the deferential lines regarding public use expressed in *Berman* and *Midkiff*:

> Two polar propositions are perfectly clear. On the one hand, it has long been accepted that the sovereign may not take the property of *A* for the sole purpose of transferring it to another private party *B*, even though *A* is paid just compensation. On the other hand, it is equally clear that a State may transfer property from one private party to another if future "use by the public" is the purpose of the taking; the condemnation of land for a railroad with common-carrier duties is a familiar example. Neither of these propositions, however, determines the disposition of this case.

In affirming the City's use of eminent domain for property acquisition as consistent with precedent, Justice Stevens emphasized the adoption of an urban renewal plan for redevelopment as negating the first proposition – the redeveloper would be carrying out a public use, i.e. the prospect of community-wide benefit through enticing an outside entrepreneur who would create jobs, which was a "public use" because it had a "public purpose". The decision in this case might be criticized for conflating "public purpose" with the constitutional requirement of "public use". Justice Stevens directly addressed the issue, saying:

> [T]his is not a case in which the City is planning to open the condemned land – at least not in its entirety – to use by the general public. … But although such a projected use would be sufficient to satisfy the public use requirement, this "Court long ago rejected any literal requirement that condemned property be put into use for the general public." … Indeed, while many state courts in the mid-19th century endorsed "use by the public" as the proper definition of public use, that narrow view steadily eroded over time. Not only was the "use by the public" test difficult to administer (*e.g.*, what proportion of the public need have access to the property? at what price?), but it proved to be impractical given the diverse and always evolving needs of society. Accordingly, when this Court began applying the Fifth Amendment to the States at the close of the 19th century, it embraced the broader and more natural interpretation of public use as "public purpose".
> (545 U.S. at 478–479 [footnotes omitted])

Once the identification of the constitutional term "public use" was satisfied by the ultimate use serving a "public purpose", the majority deferred to the local policy judgment and affirmed the taking.

> Viewed as a whole, our jurisprudence has recognized that the needs of society have varied between different parts of the Nation, just as they have

> evolved over time in response to changed circumstances. Our earliest cases in particular embodied a strong theme of federalism, emphasizing the "great respect" that we owe to state legislatures and state courts in discerning local public needs. ... For more than a century, our public use jurisprudence has wisely eschewed rigid formulas and intrusive scrutiny in favor of affording legislatures broad latitude in determining what public needs justify the use of the takings power.
>
> (545 U.S. at 482–483 [footnotes omitted])

Justice Stevens underscored this deference, adding:

> Those who govern the City were not confronted with the need to remove blight in the Fort Trumbull area, but their determination that the area was sufficiently distressed to justify a program of economic rejuvenation is entitled to our deference. The City has carefully formulated an economic development plan that it believes will provide appreciable benefits to the community, including – but by no means limited to – new jobs and increased tax revenue. As with other exercises in urban planning and development, the City is endeavoring to coordinate a variety of commercial, residential, and recreational uses of land, with the hope that they will form a whole greater than the sum of its parts. To effectuate this plan, the City has invoked a state statute that specifically authorizes the use of eminent domain to promote economic development. Given the comprehensive character of the plan, the thorough deliberation that preceded its adoption, and the limited scope of our review, it is appropriate for us, as it was in *Berman*, to resolve the challenges of the individual owners, not on a piecemeal basis, but rather in light of the entire plan. Because that plan unquestionably serves a public purpose, the takings challenged here satisfy the public use requirement of the Fifth Amendment.
>
> (545 U.S. at 483–484 [footnotes omitted])

Justice Kennedy concurred, but held out the possibility that there may be cases that would not survive the typical "rational basis" analysis, which defers to policy-maker classifications, but agreed that this was not such a case (545 U.S. at 490–493). Justice Thomas dissented, advocating a re-examination of the case law on "public use", suggesting that precedent on the subject had been misguided since the nineteenth century (*Id.* at 505–523).

Justice O'Connor's vigorous dissent seemed to tap a public mood of scepticism in government activity. O'Connor had been the author of *Midkiff*. O'Connor distinguished *Berman* and *Midkiff*, finding that they remedied a harm, rather than advancing a public good, and finding that distinction crucial:

> The Court's holdings in *Berman* and *Midkiff* were true to the principle underlying the Public Use Clause. In both those cases, the extraordinary, precondemnation use of the targeted property inflicted affirmative harm on

> society – in *Berman* through blight resulting from extreme poverty and in *Midkiff* through oligopoly resulting from extreme wealth. And in both cases, the relevant legislative body had found that eliminating the existing property use was necessary to remedy the harm. … Thus a public purpose was realized when the harmful use was eliminated. Because each taking *directly* achieved a public benefit, it did not matter that the property was turned over to private use. Here, in contrast, New London does not claim that Susette Kelo's and Wilhelmina Dery's well-maintained homes are the source of any social harm. Indeed, it could not so claim without adopting the absurd argument that any single-family home that might be razed to make way for an apartment building, or any church that might be replaced with a retail store, or any small business that might be more lucrative if it were instead part of a national franchise, is inherently harmful to society and thus within the government's power to condemn.
>
> (545 U.S. at 500–501)

In a memorable passage, O'Connor took issue with Justice Kennedy's position that a court might use the Due Process clause to strike down an unreasonable use of eminent domain when that exercise has no rational basis: "The specter of condemnation hangs over all property. Nothing is to prevent the State from replacing any Motel 6 with a Ritz-Carlton, any home with a shopping mall, or any farm with a factory" (*Id.*). No subsequent Supreme Court case has limited or overruled *Kelo*; however, the margin of the vote may foreshadow additional court forays into the area.

Self-imposed state limitations on compulsory acquisition

The result in *Kelo* set off a firestorm of public controversy. While the Court's decision allowed for the use of eminent domain for economic development, Justice Stevens' majority decision left open significant room for manoeuvre under state constitutional or statutory law to limit the power of eminent domain:

> In affirming the City's authority to take petitioners' properties, we do not minimize the hardship that condemnations may entail, notwithstanding the payment of just compensation. We emphasize that nothing in our opinion precludes any State from placing further restrictions on its exercise of the takings power. Indeed, many States already impose "public use" requirements that are stricter than the federal baseline. Some of these requirements have been established as a matter of state constitutional law, while others are expressed in state eminent domain statutes that carefully limit the grounds upon which takings may be exercised. As the submissions of the parties and their *amici* make clear, the necessity and wisdom of using eminent domain to promote economic development are certainly matters of legitimate public debate. This Court's authority, however, extends only

> to determining whether the City's proposed condemnations are for a "public use" within the meaning of the Fifth Amendment to the Federal Constitution. Because over a century of our case law interpreting that provision dictates an affirmative answer to that question, we may not grant petitioners the relief that they seek.
>
> (545 U.S. at 489 [footnotes omitted])

While the Supreme Court upheld the use of the eminent domain power for economic development, the political backlash at the appropriation of Kelo's "little pink house" was overwhelming. The Castle Coalition,[7] self-described as "Citizens fighting eminent domain abuse", reported the results of that backlash in 2010:

Simply put, the backlash to *Kelo* has been unprecedented. In the past five years:

- **9** state high courts have limited eminent domain powers
- **43** state legislatures have passed greater property rights protections
- **44** eminent domain abuse projects have been defeated by grassroots activists
- **88** percent of the public now believe that property rights are as important as free speech and freedom of religion (Ewing, 2010).

The Coalition released a "report card" that graded states on their efforts to "reform" eminent domain (Castle Coalition, 2007). While purporting not to prevent "traditional" use of eminent domain for public works projects, such as roads, libraries and sewage treatment plants, the report card takes a negative view of the use of eminent domain for lands that would, for example, be sold or otherwise used for social projects. Because the bulk of eminent domain cases are at the state and local level, it matters greatly what the limits of that power would be, especially for residential projects that increase density.[8]

Post-Kelo limitations on densification through eminent domain

Property rights advocates suggest two sets of changes to current eminent domain practice: the definition and application of "blight" and prohibition of eminent domain for economic development. Both have a profound effect on increasing density through land acquisition. Let us examine each.

Blight

Since *Berman v. Parker*, the presence of unblighted property in the midst of a blighted area has not been a reason to prevent a public agency from acquiring the entire area with a view toward redevelopment, either by itself or through contract with a private developer. This view was reaffirmed in *Kelo*, albeit by a bare majority of the Supreme Court. A number of states, prompted by Justice Steven's acknowledgement that they could have a higher bar for the use of eminent domain, amended either their state constitutions or statutes to do just that. Some states required that blight be dealt with on a parcel-by-parcel basis. This group includes Alabama, Kansas, Michigan, Oregon, South Carolina, Wisconsin and West Virginia (Castle Coalition, 2007). Others changed the definition to make it more difficult or impossible to acquire private lands through an eminent domain decision based on blight. This group includes Florida (which directs local government to use its powers to abate nuisances instead of land acquisition), Michigan (which requires "clear and convincing evidence" to support a blight designation), Minnesota and Arizona (which require an actual danger to public health or safety and no feasible alternative to the use of eminent domain in a blighted area) and New Mexico (which prohibits use of eminent domain based on blight) (*Id.*). Still others have revised the definition of blight to be more specific. This group includes Georgia, Tennessee, Virginia and Wisconsin (*Id.*).

The practical impact of these restrictions is to limit or prevent land assembly that may be necessary for housing projects, either by themselves or as a part of a "mixed use development" with other uses. The owners of holdout parcels have a significantly enhanced bargaining position than their parcels would otherwise fetch in the marketplace (although it may act as a "reality check" to ambitious redevelopment schemes). The holdout problem may be an especially acute problem in those "rust belt" cities that have declining populations. Under such restrictions, the needed urban renewal that was found valid in *Berman* would have been much more expensive and time-consuming, if it occurred at all.

Restrictions on purposes of acquisition

Other reactions to *Kelo* were restrictions or prohibitions on the purposes for which eminent domain could be used. Some states prohibited eminent domain to be used for economic development. This group includes Alabama, Louisiana, Missouri, North Carolina, North Dakota, Nebraska, New Hampshire and South Carolina. However, there are interesting exclusions from these limitations. Louisiana excludes use of the power for industrial parks, while the prohibition in Missouri and Nebraska is only to the use of eminent domain "primarily" for economic development in the former and "solely" in the latter. Alabama has perhaps the most stringent prohibition, providing that eminent domain could not be used for "private retail, office, commercial, industrial, or residential development; or primarily for enhancement of tax revenue; or for transfer to a person, nongovernmental entity, public-private partnership, corporation, or other business entity" (Castle Coalition, 2007).

Some states (Arizona and Virginia) require that property acquired by eminent domain be used by the public (Castle Coalition, 2007). Florida imposes a ten-year waiting period on transfer of lands acquired by eminent domain to private persons or entities (*Id.*). Others allow such transfers if voted upon by elected officials, including Florida (which requires a three-fifths majority of both houses of the legislature) and Georgia (*Id.*). Finally, there are related prohibitions on transfers of land acquired through eminent domain to private persons. This group includes Kansas, Illinois, Oregon and South Dakota and note that Texas requires such acquisitions not be "pretextual" and private use is not its main objective (*Id.*).

These restrictions are even more significant than the limitations on declaring a parcel or area to be blighted. For to some advocates and ideologues that favour limited government and the private sector they fulfil a heartfelt need, but do so at the risks of impoverished cities and social unrest. Moreover they enhance the position of the holdout and make publicly led redevelopment more difficult or impossible. To be fair, there were and are real abuses of eminent domain, which the *Kelo* controversy has made the centre of public debate (Somin, 2015). Whether courts will do any better than elected officials, however, is open to debate.

Other Post-Kelo restrictions

Some restrictions did not emanate directly from *Kelo*, but nevertheless have a significant impact on the potential for the use of eminent domain for densification. Some states provide for a "right of first refusal" in the event the purpose for which the eminent domain action taken was not realized. This group includes Louisiana, South Dakota and Wyoming. A Connecticut legislative report also includes Alabama, Florida and Minnesota (Castle Coalition, 2007 and Ohio Legislative Research Report, 2006). These rights are usually funded by the public agency that authorized the condemnation. Some states, notably Indiana and Minnesota, have also legislated an increment to be awarded over and above just compensation, thus giving landowners extra compensation (Ohio Legislative Research Report, 2006). This increment may include the award of a "transferable development right", which may allow the landowner to sell the right to additional density on other lands on the market (Higgins, 2008). Transferable development rights may be used to transfer possible density from environmentally constrained areas or to allow for increased density. While still in its beginning stage, their potential for increasing density is significant.

Others have found *Kelo*-like restrictions in their own state constitutions, e.g. to limit the use of eminent domain for economic development.[9] Again, the implications of such limitations may be profound and may be changed only as the state constitution (or its interpretation) may change.

While it would be a mistake to equate all housing (and all densification) as included under "economic development", the similarities between the two are striking, especially if the public agency does not undertake the housing itself or through another public agency, such as a housing or urban renewal authority. More typically, the public agency will acquire the land and dispose of it under a development agreement with a private party. To make such agreements economically attractive to developers, local governments will provide for "mixed use" zoning or commercial zoning on part of the subject property. Relationships of public bodies, especially urban renewal agencies, and the private sector run the gamut. Many public bodies advertise "requests for proposals" and hold "beauty contests" to determine which private developer with which it would be most comfortable to enter into a development agreement. Since in most states, such agreements are not subject to a public bidding process that would seek the lowest responsible bidder for the acquisition of a commodity, there is a great deal of subjectivity (and possible room for corruption) in the award of such an agreement, an outcome that has not helped the reputation of redevelopment agencies. Even without the tinge of corruption, large and politically powerful developers tend to play an inordinately large part in the politics of redevelopment.

In some states, these are the very outcomes that would be prohibited.

Conclusions

Eminent domain is just one of the tools that a state or local government may use to increase density. Other tools include the use of "density bonuses" for developers who fulfil various local planning objectives, including outright authorization for increased density under local planning and zoning, enactments, a bonus for increased density in certain designated areas or provision of affordable housing (Cohen, 2015), inclusionary zoning to require that certain developments provide a part of that development to be open for low and moderate income housing. Inclusionary zoning may be defined as "a land use practice that requires affordable housing units to be built along with market rate housing, typically in new developments" (Housing Land Advocates, 2014).

However, these alternatives depend largely on the goodwill of the private sector and may not allow for the timelessness and flexibility of direct public action. Eminent domain provides for a more immediate response to a need for densification by allowing land to be acquired immediately with just compensation and subject to construction by the agency itself or through a development agreement. However, as we have seen, these results depend on the laws of each state.

For purposes of comparison, let us review aspects of the system of compulsory purchase (eminent domain) in the United States against the list of questions this book poses regarding its use for densification purposes:

1 Under what circumstances is the taking of property rights via compulsory acquisition to increase densification justified in practice and in theory?

The answer in the United States is "it depends", because some states limit or prohibit use of eminent domain for uses that are not classically governmental in nature, such as schools, libraries or road widening. While the Supreme Court does not prohibit use of condemnation for these purposes, it is essential that the prudent land owner and public official consult the state constitution and laws, as well as the case law and practice of a state before taking action.

2 What specific public benefits are given for compulsory acquisition?

With the same caveat regarding the federal-state distinctions, the response will depend on any state limitations placed on the public use aspect of the eminent domain action. *Kelo* demonstrates that federal courts, at least for the moment, are deferential to assertions of public use. Otherwise, the legal authorities in the several states must be consulted.

3 What role does the development industry play in facilitating, encouraging or promoting compulsory acquisition?

The development industry (e.g. realtors, industrial and commercial associations) plays a political role in compulsory acquisition in the same way that neighbourhood associations, minority group representatives and other interests do. Each of these groups may also participate in litigation to deal with the legal propriety of individual acquisitions, but such participation is largely voluntary.

4 What compensation or offsets are offered for such property acquisition, and how are they funded/provided? What is the value basis for compensation? Are some land parcels resumed without compensation?

Universally in the United States, fair market value is the determinant of just compensation, which is constitutionally required. The measure of just compensation varies with the approach taken by appraisers and courts. Most eminent domain cases are settled, rather than tried, and settlement agreements may contain a number of incentives given to the landowner that would not otherwise be considered at trial. If the former use of a parcel is resumed, the owner would still be entitled to compensation for the interference with use for the period of interruption.

5 Does compulsory acquisition require special legislation or planning controls, or is it constitutionally allowed? What appeal rights are allowed? What obligation is there to negotiate?

Compulsory acquisition is viewed in the United States as an attribute of sovereignty and thus does not require special legislation, although special legislation may be used to provide limits to its use. Depending on the state, there may be an obligation to negotiate both before and during an eminent domain proceedings and eminent domain actions may be required to conform to comprehensive planning policies. Because eminent domain is constitutionally based, there is always an opportunity to challenge its exercise.

6 Is there a local or national history of compulsory property acquisition by government for a range of purposes?

As noted above, the response to both questions is affirmative.

7 Is compulsory acquisition restricted to certain types or locations of densification?

Again, this will depend on the state. While there is no known legislation on the limit of the type or location of potential densification, legislative bodies may make that policy determination, which courts will honour.

8 Where existing housing is acquired, is there any obligation to provide alternative housing in lieu?

While the federal government may require relocation assistance for federally funded projects, each state makes its own determination as to state and local projects.

9 Are other methods also used to directly bring about densification? If so, when is each method (compulsory acquisition and other method(s)) used, and how are choices between them made?

There are other methods, as mentioned above, including provision of an additional increment of compensation to some (e.g. landowners who reside on the land taken) or all condemnees, the grant of a transferable development right that may be used elsewhere and sold in the market and a right of first refusal for the return of unused land previously taken. Once again, the provision of such benefits to encourage densification is a prerogative of the individual public agency.

Notes

1 Substantive Due Process was a legal theory by which courts judged the fairness and efficacy of substantive legislation, as opposed to the manner in which that legislation was applied. This theory was dominant from the 1870s through about 1940 at the federal level, and is still influential in many state courts (Chemerinsky, 1999).
2 *See Barron v. Baltimore*, (32 U.S. 243 [1833]). Justice Marshall, a participant in the Constitutional Convention and aware of the history of the adoption of the Bill of Rights, declared for the majority that the first eight amendments to the constitution were restraints only against the federal government and that landowners must look to state law to deal with harm wrought by state or local governments.
3 The Institute for Justice represented Kelo. That entity often champions libertarian and property rights causes. See http://ij.org/.
4 See *Berman v. Parker*, (348 U.S. 26 [1954]) and *Hawai'i Housing Authority v. Midkiff*, (467 U.S. 229 [1984]).
5 *Berman v. Parker*, (348 U.S. 26 [1954]).
6 *Hawai'i Housing Authority v. Midkiff*, (467 U.S. 229 [1984]).
7 The Coalition describes itself as a grassroots project of the Institute for Justice, the public interest law firm that represented Kelo. http://ij.org/report/50-state-report-card/.
8 Under the Uniform Relocation Assistance and Land Acquisition Policies Act of 1970, (42. U.S.C. Ch. 61), relocation assistance to residents and businesses is required if a project receives federal funds. See https://portal.hud.gov/hudportal/HUD?src=/program_offices/comm_planning/affordablehousing/training/web/relocation/overview.

The National Conference of Commissioners on Uniform State Laws has recommended a Model Eminent Domain Code that provides for relocation assistance; however that Code is not binding and relocation assistance provisions must be enacted by individual states. See Eminent Domain Code Summary at www.uniformlaws.org/ActSummary.aspx?title=Eminent%20Domain%20Code. In practice, that assistance most often takes the form of monetary compensation, rather than government provision of new premises.

9 *See County of Wayne v. Hathcock*, (684 NW2d 765 [Mi, 2004]); *City of Norwood v. Horney*, (853 NE2d 1115 [Oh, 2005]); *Board of Commissioners of Muskogee County v. Lowery*, (136 P3d 639 [Ok, 2006]).

References

Castle Coalition (2007) See 50 State Report Card (2007) at http://ij.org/report/50-state-report-card/ (accessed 11 April 2018).

Chemerinsky E (1999) Substantive Due Process, *Touro Law Review*, 15, 1501.

Cohen R (2015) Density Bonuses: A Tool for Affordable Housing Development, *Non-profit Quarterly*, 31 July at https://nonprofitquarterly.org/2015/07/31/density-bonuses-a-tool-for-affordable-housing-development/ and Los Angeles, California Ordinance No. 179681 providing for extra density near mass transit stops at https://nonprofitquarterly.org/2015/07/31/density-bonuses-a-tool-for-affordable-housing-development/ (accessed 11 April 2018).

Ewing B (2010) Castle Coalition, *Kelo: Five Years Later* at www.breitbart.com/big-government/2010/06/25/kelo-five-years-later/ (accessed 11 April 2018).

Higgins N (2008) Transfer Development Rights, *Open Space Seattle 2100*, University of Washington at http://depts.washington.edu/open2100/pdf/3_OpenSpaceImplement/Implementation_Mechanisms/transfer_development_rights.pdf (accessed 11 April 2018).

Housing Land Advocates (2014) *Inclusionary Zoning in Oregon* at https://housinglandadvocates.org/resources/land-use-and-housing/inclusionary-zoning-in-oregon/ (accessed 11 April 2018).

Levy L (1999) *Origins of the Bill of Rights*. New Haven, CT: Yale University Press.

Ohio Legislative Research Report (2006) *Recently Adopted Eminent Domain Laws* at www.cga.ct.gov/2006/rpt/2006-R-0394.htm (accessed 11 April 2018).

Somin I (2015) My New Book "The Grasping Hand: Kelo v. City of New London and the Limits of Eminent Domain", *Washington Post*, 20 May at www.washingtonpost.com/news/volokh-conspiracy/wp/2015/05/20/my-new-book-the-grasping-hand-kelo-v-city-of-new-london-and-the-limits-of-eminent-domain/?utm_term=.e24204db82b9 (accessed 11 April 2018).

University of Missouri at Kansas City (n.d.) *Exploring Constitutional Conflicts: The Incorporation Debate* at http://law2.umkc.edu/faculty/projects/ftrials/conlaw/incorp.htm (accessed 11 April 2018).

4 Compulsory acquisition in UK public housing estate renewal

Legal, planning and project delivery perspectives

Gary Cox

Introduction

Comprehensive statutory powers for the compulsory acquisition of land and property by public authorities have existed in England and Wales since the mid-nineteenth century.[1] The law is complex and resides in many different statutes enacted at various times, often with provisions buried in omnibus legislation. There is no single compulsory purchase code, despite attempts over the years to reform and consolidate this area of law (Law Commission, 2003; 2004). This chapter will review the evolution of the law permitting public bodies to acquire land, both compulsorily and by agreement. Then the processes and procedures pertaining to acquisition powers are described, particularly as they relate to the justification for use of compulsory powers (a compelling case in the public interest), the role of public local inquiries and the vexed issue of the assessment of the amount of compensation.

Powers of compulsory acquisition have been used extensively in the United Kingdom to facilitate large scale urban change, for instance in establishing New Towns, large scale town centre improvements, and major urban regeneration schemes on brownfield or previously developed land. Use of such enabling powers has been essential in securing land assembly and vacant possession for highly complex transformative urban redevelopment schemes. The real-life examples presented at the end of this chapter examine the special case of public housing estate renewal schemes in the UK, the majority of which involve redevelopment to higher densities. These large post-World War II public housing estates were initially planned as mixed communities but over the decades developed a range of physical, environmental, social and economic problems. Estate renewal schemes emerged over the past few decades as a form of public private partnership, which at once facilitated the renewal of the estates with private finance together with achieving increasing densities and social mix. However, there are numerous practical and legal obstacles to overcome in order to achieve vacant possession of an estate ear-marked for comprehensive or partial renewal. This is particularly the case due to the fact that many properties on public housing estates are now privately owned as a result of the Right to Buy policy initiated through the *Housing Act 1980*. Estate renewal projects perhaps demonstrate the challenges of compulsory acquisition most clearly. However, the legal processes of compulsory purchase, particularly the local public inquiry, can focus project sponsors on the underlying project justifications and rationale, which can be insufficiently developed in a purely design-led process.

A brief history of compulsory property acquisition

The origins of contemporary compulsory purchase powers have their roots in the early nineteenth century. These powers, developed over the course of the century, were put in place to promote the industrial revolution. The totemic industrial purpose in the nineteenth century was the expansion of railways. The earliest Act of Parliament authorising the compulsory purchase of specific land for a railway was the *Wandsworth to Croydon Railway Act* of 1801. This legislation both set out procedures for acquiring the land and the mechanisms for providing compensation. The use of such powers multiplied in the mid-nineteenth century to include reservoirs, sewerage systems, street widening but also for the defence of the realm (Barnes, 2014). Using individual enactments soon led to complications and inconsistencies in application and interpretation. Hence, a series of Acts were passed mid-century to consolidate standard provisions in over-arching legislation. These were the *Lands Clauses Consolidation Act 1845*, the *Railways Clauses Consolidation Act 1845*, the *Waterworks Clauses Act 1847* and the *Cemeteries Clauses Act 1847* (Barnes, 2014). Collectively, these Acts were referred to as the Land Clauses Acts. Fundamentally this legislative framework of standard provisions in foundational legislation supported by context specific provisions in special legislation continues to this day. Though the Land Clauses Consolidation Acts were incorporated into the *Public Health Act 1875* (s.176), alongside a general power to purchase land (s.175), the newly constituted Local Government Boards did not use these powers to any extent (Cullingworth *et al.*, 2015).

A watershed moment in British urban planning came after 1945 and the cessation of hostilities in World War II. Various town planning statutes had been enacted from 1909 onwards, though initially the focus was more on housing than comprehensive planning. Generally, the pre-WWII planning arrangements were regarded as ineffective, being largely optional for local authorities and essentially regulatory in nature. A comprehensive planning 'system' was to await the enactment of the *Town and Country Planning Act 1947* by the new Labour Government of Prime Minister Clement Atlee. The Labour Manifesto of 1945, 'Let Us Face The Future', is a bold statement of the reforming intent of the future government (Pearce, 1994). The document announced an ambitious house building programme as well as action to ensure an efficient building industry. Land was seen as key with the heralding of 'a radical solution for the crippling problems of land acquisition and use in the service of the national plan' (The Labour Party, 1945, 9). A first step to full land nationalisation was wider and speedier powers for local government authorities 'to acquire land for public purposes wherever the public interest so requires' with provision for 'fair compensation' and a revenue stream for public funds from 'betterment' (The Labour Party, 1945, 9).[2]

The *New Towns Act 1946* became a model for future large scale comprehensive urban development. The New Towns as initially conceived were promoted to take pressure off the growth of London, both residential and industrial (Barlow Commission, 1940). As such, they were a mechanism for the decentralisation of London in the context of the need to reconstruct much of war-damaged Britain (Harwood, 2015). Designated New Towns were governed by a development corporation with wide powers to acquire land, either by agreement or compulsory purchase. The *New Towns Act 1946* (s.4) followed a model to be frequently repeated: that of incorporating within it the compulsory purchase provisions of existing enactments, such as the Land Clauses Acts.

In the 1950s through to the late 1970s, the evolving framework of compulsory purchase powers was used by both local and central governments to facilitate slum clearance and redevelopment, major transport infrastructure (motorways, ring roads, ports and airports) as well as town centre improvements. The next major milestone came at the start of the Thatcher era with the passage of the *Local Government, Planning and Land Act 1980*. This was put in place to enable a market-driven rather than a needs-driven approach to urban growth and regeneration. The 1980 Act enabled the establishment of urban development corporations (UDCs) with similar powers to those of the New Towns, notably the London Docklands Development Corporation. These bodies were given planning powers and the power to have land vested in them without any recourse to appeal mechanisms. As Hall (1999) observes, the underlying concept of UDCs was the leverage of private sector investment to kick-start urban regeneration.

New institutional structures were established in the 1990s to more widely initiate urban regeneration through land development. English Partnerships was set up in 1994 by the Conservative Government of John Major with a wide remit to regenerate land in England that was vacant, underused, contaminated, derelict or unsightly (Roberts and Sykes, 2000). It had similarly wide powers of land acquisition and compulsory purchase as the preceding statutory bodies. The New Labour Government in 1997 established Regional Development Agencies across England with a broad range of responsibilities ranging from economic development to urban regeneration. Again, these bodies were provided with wide ranging powers of land acquisition.[3] The *Greater London Authority Act 1999* (s.333ZA) provides for a general power for the Mayor of London to compulsorily acquire land for housing and regeneration. The model of specialist agencies being vested with wide ranging land acquisition powers was repeated for subsequent special purpose bodies such as Mayoral development corporations and for the main unified housing and regeneration body in England, Homes England.

Compulsory acquisition for public purposes: legal pathways

In English law, the ultimate or 'radical' title to all land resides in the Crown (as the state). The fundamental concept of ownership is 'estates' relating to that radical title. The doctrine of estates is complicated by the notion that the property owner merely owns not the land itself but a 'slice of time' in the land. In the English system, there is a major distinction between freehold land and leasehold land (Gray and Gray, 2009). The former implies an estate of unlimited duration and the latter specifies the time period that exclusive possession is granted. Common leases are 125 years, 250 years and 999 years.

Unlike most other jurisdictions covered in this book, the United Kingdom does not have a written constitution. Instead, constitutional conventions have evolved over hundreds of years of history. In the British constitution, Parliament is sovereign, hence powers relating to compulsory acquisition are derived from legislation and not the constitution. Use of the so-called prerogative powers of the executive to acquire private property without compensation has been found unconstitutional by the Appellate Committee of the House of Lords (now called the UK Supreme Court) (Loveland, 2018). The entirety of the law of compulsory acquisition and the law of assessment are statute based and there is no common law right or extant prerogative power permitting the taking of property by the state without the explicit authority of Parliament (Barnes, 2014).

Having established that the sole route for use of compulsory acquisition powers is statute, what are the key sources of that power? Powers of compulsory purchase are to be found in a large number of statutes.[4] Here of major relevance are the powers vested in local government authorities or 'principal councils'. These are contained in the *Local Government Act 1972* (ss.120–123). There are wide powers of acquisition by agreement for any purpose authorised by legislation or for 'the benefit, improvement or development of their area' (s.120(1)). Similarly, local authorities are authorised to acquire land through compulsory purchase 'whether situated inside or outside their area' – so the general power is wide and unrestricted in its geographical scope. However, for urban development projects, such as estate renewal schemes or town centre improvements, local authorities will more often rely on powers contained in planning legislation. Land acquisition powers are set out in Part IX of the *Town and Country Planning Act 1990* and the relevant provisions should be read together. Section 226 establishes the key power of compulsory purpose but only 'if the authority think that the acquisition will facilitate the carrying out of development, re-development or improvement on or in relation to the land'. However, this purpose is not sufficient in and of itself. The local authority must establish that the development is likely to improve or promote the economic, social or environmental well-being of its area (s.226(1A)). This test forms a central part of a local authority's statement of reasons or statement of case accompanying the application to the Secretary of State for the use of compulsory purchase powers. This statement will also delineate the order land, its location and current use.

The above discussion relates to sources of powers to compulsorily purchase land by public bodies, where the purpose relates to those bodies' statutory functions or for a more widely drawn concept of public benefit. The former purpose could and sometimes can lead to a situation where a public body's powers are used for purposes tangential to a direct public benefit. Procedures for acquisition reside in separate enactments which have similarly evolved over time since the Land Clauses Acts in the mid-nineteenth century. These procedures may be regarded as to effect good public administration. But as Barnes (2014) points out, these procedures act as safeguards for property owners that the taking of their land is justified and required and also that fair compensation is paid for the loss of the benefit of that land. The basic procedures for the granting of a compulsory purchase order are contained in the *Acquisition of Land Act 1981*. These procedures provide for an 'acquiring authority' (say, the local authority) to apply to the 'confirming authority' (usually the Secretary of State for Housing, Communities and Local Government) for the grant of a compulsory purchase order (s.7(1)).

Any person can make an objection to a compulsory purchase order and there are a number of grounds that can sustain an objection (Denyer-Green, 2014). A particular set of grounds relates to rights conferred by the European Convention on Human Rights, which has been transposed into UK law through the *Human Rights Act 1998*. However, if the issue relates exclusively to the amount of compensation to be paid, the matter needs to be referred to the Upper Tribunal (Lands Chamber). It is often the practice for an acquiring authority to negotiate with an objector but, should objections remain, the Secretary of State will hold 'public local inquiries as are directed by this Act and such other public local inquiries as he may think fit' (*Acquisition of Land Act 1981*, s.5(1)). The inquiry is conducted by a member of the Planning Inspectorate according to statutory rules of procedure (*The Compulsory Purchase (Inquiries Procedure) Rules 2007*). Such an inquiry can significantly delay development projects as the entire inquiry and reporting process can around two years to complete.

After considering the inspector's report, the Secretary of State will then formally decide to confirm, modify or reject the application for the compulsory purchase order. Although the Secretary of State can make a decision which is contrary to the inspector's recommendations, this is uncommon. There is a statutory right of appeal. Any aggrieved person has six weeks to apply to challenge the validity of the compulsory purchase order in the High Court on the grounds that the order has been made contrary to the relevant legislative powers or that there has been a failure to comply with a procedural requirement (*Acquisition of Land Act 1981*, s.23 & s.24(2)). The result if the objection is upheld will be the administrative law remedy of quashing the order. Thus, this appeal route is akin to judicial review. This appeal route is usually the one taken by objectors. For local government authorities as acquiring authorities, the route taken to counter the decision will usually be a claim for judicial review with potentially a wider scope of grounds for review. This claim must be brought promptly and in any event within three months of the decision. Cases are heard by the Planning Court, which is a specialist court within the High Court.

The final part of the compulsory acquisition process is the actual acquisition of rights by the acquiring authority. Basically, this can take two forms: either what is called a general vesting declaration or a notice to treat/notice of entry procedure. The latter procedure is where notices are sent to individual property owners giving them a defined period within which to respond with a notice of claim. Following the notice to treat, the acquiring authority can take possession of the land following service of a notice of entry. The acquiring authority has three years within which to exercise these rights (*Compulsory Purchase Act 1965*, s.4).

For large schemes with many interests to acquire, the above process can be cumbersome. So, an expedited route is provided via a procedure known as a general vesting declaration (*Compulsory Purchase (Vesting Declarations) Act 1981*). In stark contrast to the notice to treat procedure, the use of a general vesting declaration replaces the entire process with individual property owners with one procedure which at a set date vests the entirety of the subject land with the acquiring authority. Once the vesting declaration has been made, the procedures of the *Compulsory Purchase Act 1965* come into effect in terms of individual property owners making claims and having recourse to the Upper Tribunal (Lands Chamber). The making of a general vesting declaration can be challenged through judicial review proceedings (Denyer-Green, 2014). A high-profile example of the use of a general vesting declaration was the London 2012 Olympic and Paralympic Games. In this case, the powers of compulsory acquisition resided in the London Development Agency (*Regional Development Agencies Act 1998*, s.20(1)). The majority of the land and property required for the construction of the Olympic Park and stadiums at Stratford in East London were acquired on or before 2 July 2007 by the use of general vesting declarations (Culture, Media and Sport Committee, 2007; Bernstock, 2014).

Negotiation and compensation

This section will examine the principles relating to the compensation for land acquired both via compulsory acquisition and through acquisition by agreement. There is now an explicit obligation on acquiring authorities to negotiate with property owners prior to a public body formally deciding to use compulsory acquisition powers. Unlike the situation of federally funded projects in the United States, there is no statutory obligation to demonstrate that affected individuals and businesses have been negotiated with in good faith. However, since 2004 government guidance has indicated that compulsory purchase should be used as 'a last resort' and only where there is a 'compelling case in the public interest' (Office of the Deputy Prime Minister, 2004, §17 & §24). This has been reiterated in more recent guidance with the explicit statement that the acquiring authority should demonstrate that they have taken 'reasonable steps' to acquire the land by agreement (Department for Communities and Local Government, 2015, 6). This is a matter that will be examined by an inspector in a local public inquiry. It is now the practice for local government authorities to maintain detailed case records of discussions with local property owners prior to and after the decision to apply for a compulsory purchase order, particularly for large scale regeneration projects. These so-called informal negotiations may indeed be formalised in advance through explicit policies of the acquiring authority and form a critical part of estate regeneration schemes, as will be discussed in the next section. Given that the inquiry process may be both lengthy and costly, it is often in the interests of the acquiring authority to negotiate with property owners early in the anticipation that this may limit the scope of the compulsory purchase order. Such negotiations can continue during the local public inquiry process and even prior to notices to treat being issued. The two routes to acquisition through compulsion or by agreement, seemingly separate in the legislation, are generally employed together to achieve the outcome of land assembly and vacant possession.

The most contentious issue in compulsory land acquisition is compensation and, fundamentally, compensation for the value of the land acquired. Unsurprisingly, this is a complex area of law and valuation. The fundamental principles are seemingly simple, though their application in any given case can be fraught. These principles have been transposed into the six rules for assessing compensation in the *Land Compensation Act 1961* (s.5). These rules have evolved over time from the *Lands Clauses Consolidation Act 1845* (Barnes, 2014). The primary principle is that of equivalence. That is the compensation payable to a person who is deprived of land in the public interest should be equal to the true and actual loss suffered by that person, no more and no less. This principle is similar to that of damages in tort law. Transgressing the principle would on the one hand impose a loss on the landowner or on the other hand place an unfair burden on the public authority. For the avoidance of doubt, this touchstone principle eschews resumption without compensation. Against a strict interpretation of equivalence, the *Land Compensation Act 1973* (s.30(1)) includes additional payments called home loss payments where a person is displaced from their place of residence. This amounts to an additional 10 per cent of the market valuation up to an upper limit prescribed in regulations.[5] For non-residential property, this additional amount is 7.5 per cent of the market valuation of the property. For residential property, additional disturbance payments apply for removal expenses and the like.

The second main principle is that of open market value. This is stated in rule (2) of section 5 of the *Land and Compensation Act 1961*. This is that the compensation should reflect the value of the land if sold in the open market by a willing seller to a willing buyer. In practice, this principle creates a number of problems. The valuation date becomes a critical variable as markets and external conditions change over time. Some projects have long gestation times during which market values may become depressed 'in the shadow' of a compulsory purchase order. As a general rule, the valuation date is either the assessment date (if acquisition is by agreement) or for the 'notice to treat' procedure the date of actual entry by the acquiring authority. For the general vesting procedure, the vesting date specified in the declaration is the valuation date.

Divining the actual market for the land and what amount prospective buyers might be willing to pay is a matter of technical judgement open to dispute. The most favoured method of valuation in this context is to make comparisons with actual sales but establishing direct comparisons can be fraught with difficulties even if there is a current record of potential like sales that can be referenced. The closer to the date of compulsory acquisition the likelihood is that there will no longer be a 'willing buyer' and true comparators will be less available.

The third main principle is the value to the owner principle, which is sometimes called the *Pointe Gourde* principle from a 1947 Privy Council case where the principle was applied. This states that the compensation for the compulsory acquisition of land cannot include any increase or decrease in value of the land which is due to the scheme underlying the acquisition. The value to the owner basis for compensation goes back to the *Lands Clauses Consolidation Act 1845* (s.63). The *Pointe Gourde* case established that the compensation for the acquisition of land cannot include an increase in value which is entirely due to the underlying scheme itself. The legal principle was restated in 2004 by the House of Lords in *Waters v Welsh Development Agency*. The judgment stated that: 'It is necessary to separate from the market value of land any enhancement in value attributable solely to the presence of the acquiring authority in the market as a purchaser of the land in exercise of its statutory powers' (§21). The judgment also specifically stated that such valuation should disregard the 'special want' of the acquiring authority. Consequently, the valuation must therefore be conducted in a 'no-scheme world' (Law Commission, 2001). Again, as with open market value, this principle presents its own technical difficulties. In summary, the current position is fairly explicitly stated in a number of legal cases that the value to the owner principle is no longer appropriate, and market value plus disturbance generally should form the basis of compensation (Barnes, 2014; Denyer-Green, 2014).

Estate regeneration, land assembly and densification

National policy

Since the mid-1990s, there has been a marked shift in national policy to increase the supply of housing, especially affordable housing, through the development of brownfield sites. Developers have traditionally been reluctant to build on brownfield sites, particularly those in central areas. Major obstacles to development have been the costs and time associated with land assembly, issues of site contamination and remediation, and also planning controls (Tallon, 2010). In 1998, the government announced a national target for at least 60 per cent of all new housing to be built on 'previously developed land' (Department of the Environment, Transport and the Regions, 2000). By 2008, this target had been achieved and it has been maintained or exceeded since then. In 2012, the government published its National Planning Policy Framework. This document sets out the government's requirements for the planning system in England.

The document sets out 12 core planning principles, one of which aims to 'encourage the effective use of land by reusing land that has been previously developed (brownfield land), provided that it is not of high environmental value' (Department for Communities and Local Government, 2012, §17). The framework also advocates the planning of larger scale sites to best achieve the supply of new housing. There is now no specific target for brownfield development.

The opportunity

Inner London and the inner areas of major British cities are typically made up of large public housing estates, some constructed in the inter-War years with the majority constructed after 1945 (van Kempen *et al.*, 2005). Typically for inner London councils, around 25–30 per cent of land is occupied by public housing (Adonis and Davies, 2015). These estates have largely remained unaltered since they were originally built, apart from modernisation to individual homes and major maintenance works to large housing blocks. A report by the Centre for London estimates that densification of existing large public housing estates has the potential to add 80,000–160,000 homes to London, which translates to 4,000–8,000 homes per year (Hanna *et al.*, 2016). A number of features of these estates make them attractive for urban regeneration schemes. First, these estates are often well-located and sometimes exist alongside expensive housing. Second, they were often built at low densities, despite the tower blocks that may be the dominant architectural feature. Often given public funding constraints, they are difficult to improve without some form of private sector involvement.

Current estate renewal practice

In January 2016, the then Prime Minister David Cameron announced a £140 million fund to transform the country's most run-down public housing estates, replacing them with new attractive and safe housing. The government's policy objectives were twofold: to increase housing supply through densification, and to reverse social deprivation by improving the life chances of the most disadvantaged residents of these areas (Cameron, 2016). Estate renewal covers a variety of projects delivered in different ways (McLaughlin, 2015). There are small scale infill projects taking up the opportunities of under-used sites on estates, for example car parks or garages. Other projects are long term commitments and entail comprehensive transformation of entire estates. The larger scale schemes generally include renewal of social infrastructure, such as libraries and community centres, town centre upgrades and public realm improvements. Transport enhancements are often crucial elements of these projects. Such projects are usually conducted as partnerships with private sector developers or house builders, either contractually through a development agreement or through a joint venture company. The lead partner and project initiator will be either a local council or a non-profit housing association. Peabody, a housing association founded in the mid-nineteenth century, is engaged in the regeneration of a 1930s estate at St John's Hill in Battersea in inner London and in an ambitious scheme to regenerate the extensive Thamesmead Estate in south east London built in the late 1960s.

Large scale estate regeneration schemes can take many years to complete, with ten years or more not being untypical. With constraints on local government finance and borrowing caps, leveraging private finance and risk management is essential to enable these projects to commence. There is also the challenge faced by local government authorities to achieve 'best consideration' when disposing of or pooling land (*Local Government Act 1972*, s.123(2)). Hence, schemes are formulated that include a mix of tenures with a significant element of increased density and private sale and sometimes private market rental properties. This is seen as a means to maximise the effective use of land while also meeting affordable housing obligations. However, this part of estate regeneration is the most misunderstood and has led to claims of 'social cleansing' or 'stealth gentrification' (Architects for Social Housing, 2017). To enable a large estate regeneration project to be viable, often substantial public funds are necessary, either in the form of direct subsidy or equity share by the local authority partner, forward purchase agreements, or through grants from central government or the Greater London Authority. As recognised by recent government guidance, financing of estate regeneration is a significant challenge particularly in the context of changing risk profiles through the life of a project. Consequently, testing for financial viability is an on-going issue throughout the project cycle and a major consideration in the early stages of project preparation (Department for Communities and Local Government, 2016a).

Process of estate regeneration – planning frameworks

Concurrently with financial modelling, most projects conduct some form of master planning exercise. The proposals for the regeneration of the Heygate Estate near Elephant and Castle in South London had a very long gestation period. The estate was constructed between 1971 and 1974 in a series of slab blocks connected by overhead walkways. It comprised 1,194 residential properties of which 1,015 were council tenancies and 179 had been purchased under Right to Buy. Southwark Council had been seeking to regenerate the area since 1997 and conducted an options appraisal exercise with residents in 1997–1998. A regeneration framework was prepared by the council in 2002 and subsequently a masterplan was approved in 2004. Elephant and Castle was included in the London Plan 2004 as an opportunity area with the potential for 4,200 new homes (Mayor of London, 2004). Though the initial masterplan was published in 2004, it was not until 2012 that the current framework was developed as a joint supplementary planning document and opportunity area planning framework (Mayor of London and Southwark Council, 2012). This includes the Elephant and Castle regeneration area being taken forward by Lend Lease, which comprises three development sites with a potential for around 3,000 new homes, transport improvements and new retail and community facilities. Other estate regeneration schemes use broadly similar development frameworks, sometimes in the form of an area action plan or a planning parameters document (Southwark Council, 2010; GVA Bilfinger, 2015).

Process of estate regeneration – offers to directly affected residents

The master planning process will result in the identification of an area and specific properties for future demolition. There will almost always be existing residents living in these properties. The Estate Regeneration National Strategy recognises residents as being the key partners of any regeneration scheme and that they should be given opportunities to be involved throughout the process (Department for Communities and Local Government, 2016b). Early involvement of the local community together with consistent and clear communications is a recurring theme in the current discourse around estate regeneration (London Assembly Housing Committee, 2015; Prince's Foundation and The Hyde Group, 2015; Future of London, 2016). This is the central concern of the good practice guide published by the Mayor of London. The guide emphasises that consultation is a process that should start early and continue throughout the lifetime of the project with a focus on transparent, extensive, responsive and meaningful engagement (Mayor of London, 2016).

There is a direct linkage between the development of a project master plan and a phasing and decant strategy. The term 'decant' is a slightly cumbersome term for the relocation of existing residents. This phasing and decant strategy will need to be underpinned by a resident offer policy. Some regeneration projects such as the Eastfields Estate in Mitcham, South London, condense commitments into a resident's charter (HTA *et al.*, 2016). Most large public housing estates comprise a mix of tenures: social rent (from the council or housing association), resident freeholders or leaseholders, and private tenants. The latter two categories are the result of the national Right to Buy policy. On some estates, there can be a large privately rented sector from non-resident Right to Buy landlords. Since the policy was advanced in the early 1980s, some properties have been resold numerous times. If the balance between leaseholders and social tenants is too high (say over 50 per cent), it is generally acknowledged that site assembly will be too protracted to constitute a deliverable scheme.

Resident and tenant offer policies can be complex. There is a broad duty on the acquiring authority to rehouse residential occupiers (*Land Compensation Act 1973*, s.39(1)). But there are also similarly wide powers for a local housing authority to obtain possession of properties with secure tenancies where 'the dwelling house is in an area which is the subject of a redevelopment scheme approved by the Secretary of State' (*Housing Act 1985*, Schedule 2, Ground 10A). However, possession will only be granted by a court if suitable alternative accommodation is available (*Housing Act 1985*, s.84(2)(b)). The Mayor of

London's guidance states 'that existing social housing tenants should be offered a right to return to the regenerated estate ... at the same or similar level of rent, the same level of security of tenure' (Mayor of London, 2016, 28). However, practice has varied. Southwark Council instituted a 'right to return' policy in 2007 for the Heygate Estate regeneration (Southwark Council, 2007). However, actual vacant possession was primarily secured through the development of smaller sites in the local area through housing associations. In 2007, Heygate comprised 1,212 units, of which 650 were tenanted, 442 were void and 120 were owned by leasehold interests. The aim was to secure vacant possession over a two-year period prior to any construction commencing. Acquisition of leasehold properties was progressed in parallel with the rehousing of tenants with some options for rehousing through new shared ownership or shared equity options.

The scale of the Heygate vacant possession task and the fact that it was conducted prior to construction led to claims of 'social cleansing' (Hill, 2014). Wandsworth Council in south west London has taken a different approach in its two major estate regeneration schemes – the Alton Estate in Roehampton and the Winstanley York Road Estate in Battersea. The resident offer policy was released over two years ahead of procurement of development partners for the schemes. For secure council tenants, the policy is based on a 'one-move' principle. This means that an existing tenant will have the option to move directly from their current property to a new build property in the regeneration scheme, though other options are available (Wandsworth Council, 2014a; 2015). For resident owners, whether freeholders or leaseholders, the option is a similar one-move but to a new shared equity property in the scheme. The policy also allows for 'voluntary buy-backs', meaning owners can sell their properties back to the council at any time on valuations based on the statutory principles discussed above, plus all statutory additional payments (Wandsworth Council, 2014b; 2015). Such policies are predicated on schemes being carefully phased to allow single moves to occur phase-by-phase. Even so, Wandsworth Council has needed to supply limited additional off-site housing to permit the initial phases to commence, but not to the extent undertaken by Southwark Council for the Heygate Estate.

Process of estate regeneration – compulsory purchase procedure

Preparation for the statutory procedure for compulsory acquisition generally will occur in parallel with master planning, early resident engagement and procurement of a development partner. Some projects will require a number of separate compulsory purchase orders if the entire scheme is separated into distinct temporal and spatial phases. The rationale for this is twofold. First, a notice to treat ceases to have effect three years from service of the notice (*Compulsory Purchase Act 1965*, s.5(2A)). Similarly, a general vesting declaration may only be served within three years of confirmation of a compulsory purchase order. Second, probably more significantly, the requirements of a compelling case in the public interest in the statement of reasons prepared ahead of any public local inquiry will engender a cautious approach regarding the scope of the order. This statement will need to convince the planning inspector regarding issues of funding, necessary infrastructure works, the likelihood of gaining planning permission and other matters.

An initial preparatory task, once the extent of demolition is understood through the master planning process, is to prepare a property cost estimate. This is a comprehensive estimate of the cost of all the property interests that will need to be acquired in order to gain vacant possession. Calculations are based on the statutory valuation methodologies for compulsory acquisition, together with all additional payments, land referencing costs, taxes and fees. The property cost estimate will also include legal costs for anticipated local public inquiries. This estimate will form a key input into the compulsory purchase indemnity agreement that is likely to form part of the overall legal agreement between the acquiring authority and the development partner. The indemnity agreement provides that the acquiring authority will exercise its compulsory purchase powers if and when it is proper to do so and to ultimately transfer the land to the developer. In return, the developer will pay either all or an agreed proportion of the costs incurred by the process as well as, in effect, the compensation due to the owners. These payments to the acquiring authority are usually separate to any other contractual payments payable under a development agreement or profit shares via a joint venture company.

One of the best ways to understand the process of individual estate regeneration schemes and also their planning justifications is to examine the statement of reasons/statement of case prepared by the acquiring authority prior to the local public inquiry. One example is the West Hendon Estate regeneration in North London. This is a scheme for the construction of 1,651 market sale homes and 543 affordable homes (256 social rent and 287 shared equity). The existing estate was built in the 1960s and originally comprised 680 dwellings in six blocks. The estate forms part of the Cricklewood/Brent Cross Opportunity Area which was first designated as an opportunity area in the London Plan 2004 and is now deemed to have a capacity for 10,000 new homes (Mayor of London, 2011). A compulsory purchase order was made by Barnet Council under section 226 of the *Town and Country Planning Act* 1990 on 4 June 2014. This order was for the first phase of the scheme which involved the compulsory acquisition of 34 leaseholder properties on the West Hendon Estate. A pre-inquiry meeting was held on 29 October 2014 with the inquiry formally sitting on various dates in January 2015 with a site inspection at the end of that month. At the start of the inquiry, there were 78 statutory objectors, mainly directly affected property owners, and 29 so-called non-qualifying objectors. The inspector made her report to the Secretary of State on 6 August 2015 with a recommendation for the order to be confirmed subject to some modifications. In turn, the Secretary of State confirmed the compulsory purchase order on 12 November 2015. All in all, the process took around 18 months.

A comparison between the inspector's report and the Secretary of State's confirmation letter reveals a high degree of concurrence in their assessments

(Hill, 2015). The objections related to the non-achievement of economic, social and environmental benefits, loss of open space, insufficient health and education facilities, and harm to a Site of Special Scientific Interest. In addition, objectors raised questions as to the adequacy of the consultation, the extent of meaningful negotiation with property owners, and project funding. There were also specific objections relating to individuals' personal circumstances relating to businesses and homes to be acquired. The inspector found that the purpose for which the land was being acquired was in substantial accord with the adopted planning framework for the area. Overall, the Secretary of State concluded that the 'public benefits significantly outweigh the private loss and consequently that there is a compelling case in the public interest to justify sufficiently the interference with the human rights of those individuals affected by the Order' (Secretary of State for Communities and Local Government, 2015, §27).

The planning inspector came to a very different conclusion regarding human rights in the report of the local public inquiry into the Aylesbury Estate in Southwark, South London. The Aylesbury Estate was constructed between 1966 and 1977 and comprised around 2,250 social rented homes and 500 homes in leasehold properties sold through Right to Buy. The Aylesbury Area Action Plan was adopted in 2010 and aims to regenerate the entire estate over a 15-year period (Southwark Council, 2010). The ambition of the area action plan is to replace the existing properties with 4,200 new mixed tenure homes, new schools and community facilities, and improved open space. A compulsory purchase order was made by Southwark Council under section 226(1)(a) of the *Town and Country Planning Act 1990* on 24 June 2014. This order was for the first development site which included seven blocks totalling 566 residential properties. These were to be replaced by 830 mixed tenure homes, including 424 private market homes. The inspector reported to the Secretary of State on 29 January 2016 with a recommendation for the order not to be confirmed. There were ten statutory objectors, mainly leaseholder property owners, and one non-qualifying objector. The main, though not the only, issue of contention related to the human rights of the remaining leaseholder owners. The human rights engaged in compulsory acquisition are Article 8 (right to respect for private and family life) of the European Convention on Human Rights and Article 1 (protection of property) of the Protocol to the Convention. The council's policy at the time (though changed subsequently) stated that in order to qualify for a shared equity rehousing option each household is permitted to retain only £16,000 of their capital or savings. The inspector regarded this as an intrusive level of scrutiny, particularly when compared to the situation of tenants. For some groups, such as the elderly, this policy would almost certainly have led to some leaseholders moving out of the area with the consequential impacts on family life and there were further impacts identified on ethnic minority households (Coffey, 2016). The Secretary of State agreed with the inspector that 'the interference with human rights is not proportionate in all the circumstances' (Secretary of State for Communities and Local Government, 2016, §22). Following a change of the council's policy regarding individual savings and a judicial review procedure, the original decision not to confirm the order was quashed and a new local public inquiry commenced in early 2018 to confirm the order.

Conclusion

It is commonplace for legal scholars to refer to property ownership as a 'bundle of rights' (Whelan, 1980, 119). It is now a legal convention backed by statute, that the taking of property by the state cannot occur without compensation to the original owner. As the review of the legal underpinnings to compulsory acquisition and subsequent valuation principles confirm, one of the rights in the bundle of rights is the right to fair compensation. Since the enactment of the *Human Rights Act 1998*, there is an additional human rights dimension relating to protection of property, as well as the broader consideration of respect for private and family life. This balancing of individual rights and wider societal benefits goes to the heart of the 'compelling case' that an acquiring authority must justify to the Secretary of State in order to legally exercise powers of compulsory acquisition.

The law and practice of compulsory acquisition in England and Wales in highly complex, having been constructed over a period of over 170 years both through legislation and case law. As an area of law fundamental to the operation of the modern state and the national economy, it is self-evidently in need of reform. More than a decade ago, the Law Commission, an independent body that keeps the law of England and Wales under review, issued two reports proposing fundamental reforms to compulsory purchase and the law and procedure for compensation (Law Commission, 2003; 2004). The overall recommendation was that there was a pressing need for a consolidation of the law into a compulsory purchase code. There is some recognition in UK professional and political circles of the need and rationale for reform along these lines, though such a fundamental reform is unlikely to occur in the foreseeable future.

There is currently a major issue of housing supply and delivery in the UK. Regeneration of large public housing estates has been one route to increase the supply of housing in established areas of cities and large towns. Compulsory acquisition is a key tool in enabling these highly complex projects to deliver in a timely manner but, as the examples in this chapter demonstrate, use of compulsory powers can be both technically and politically fraught. Public housing estates have a particular manifestation of land fragmentation that has been an unintended consequence of the Right to Buy policy initiated in the early 1980s – alongside social tenants, there can be substantial numbers of both resident and non-resident owners. The fundamental contention of this chapter is that compulsory acquisition powers cannot be viewed in isolation from other aspects of effective delivery of estate regeneration programmes, such as voluntary buy-back policies, commitments to rehouse existing residents in the new development, and on-going community and stakeholder engagement. The overall conclusion is that increased densities, higher quality housing and public realm, and effective delivery of additional housing can all be achieved through carefully managed upfront planning processes and continuing involvement of local residents. These residents should be provided a concrete realisable stake in the new scheme. In this way, the case for the promotion of the economic, social and environmental well-being of the area can be made in a compelling way.

Acknowledgement

The author would like to acknowledge the assistance of Virginia Blackman, Director, National Head of Compulsory Purchase, at GVA, London.

Notes

1 Note that Scotland and Northern Ireland each have separate legal systems to England and Wales, and so will not be covered in this chapter. Institutional arrangements are further complicated by the devolution of powers to the Scottish Parliament and the National Assembly for Wales that occurred in 1998.
2 A development charge was established by the *Town and Country Planning Act 1947* based on a proportion of the 'betterment' resulting from the grant of planning permission. Betterment is defined as the uplift in value of the land resulting from the 'benefit' of the planning permission (s70(2)).
3 The Regional Development Agencies were abolished following the election of the Coalition Government in 2010.
4 This chapter focuses on procedures specifically relating to urban regeneration. There are different procedures for nationally significant infrastructure projects (under the *Planning Act 2008*), projects under the *Transport and Works Act 1992*, and hybrid or specific purpose Acts, such as the *Crossrail Act 2008*.
5 The *Home Loss Payments (Prescribed Amounts) (England) Regulations 2017* prescribe a minimum payment of £6,100 and a maximum amount of £61,000.

References

Adonis A and Davies B (eds) (2015) *City Villages: More Homes, Better Communities*. London: Institute for Public Policy Research.

Architects for Social Housing (2017) *Social Housing: Demolitions, Privatisations & Social Cleansing*. Accessed 19 January 2018 at https://architectsforsocialhousing.wordpress.com/2017/05/13/social-housing-demolitions-privatisations-social-cleansing/.

Barlow Commission (1940) *See* Royal Commission on the Distribution of the Industrial Population.

Barnes M (2014) *The Law of Compulsory Purchase and Compensation*. Oxford and Portland, Oregon: Hart Publishing.

Bernstock P (2014) *Olympic Housing: A Critical Review of London 2012's Legacy*. Aldershot, Surrey: Ashgate Publishing.

Cameron D (2016) *Estate Regeneration*. London: UK Government. Accessed 22 January 2018 at www.gov.uk/government/speeches/estate-regeneration-article-by-david-cameron.

Coffey L (2016) *Application for Confirmation of the London Borough of Southwark (Aylesbury Estate Site 1B-1C) Compulsory Purchase Order 2014: CPO Report to the Secretary of State for Communities and Local Government*. London: The Planning Inspectorate.

Convention for the Protection of Human Rights and Fundamental Freedoms, opened for signature 4 November 1950, 213 UNTS 221 (entered into force 3 September 1953). Accessed 19 January 2018 at www.echr.coe.int/Documents/Convention_ENG.pdf.

Cullingworth B, Nadin V, Hart T, Davoudi S, Pendlebury J, Vigar G, Webb D and Townshend T (2015) *Town and Country Planning in the UK*. London: Routledge.

Culture, Media and Sport Committee (2007) *London 2012 Olympic Games and Paralympic Games: Funding and Legacy Second Report of Session 2006–07*. London: House of Commons.

Denyer-Green B (2014) *Compulsory Purchase and Compensation*. London: Routledge.

Department for Communities and Local Government (2012) *National Planning Policy Framework*. London: Department for Communities and Local Government. Accessed 19 January 2018 at www.gov.uk/guidance/national-planning-policy-framework.

Department for Communities and Local Government (2015) *Guidance on Compulsory Purchase Process and the Crichel Down Rules for the Disposal of Surplus Land Acquired by, or under the Threat of, Compulsion*. London: Department for Communities and Local Government.

Department for Communities and Local Government (2016a) *Estate Regeneration National Strategy: Financing and Delivering Estate Regeneration*. London: Department for Communities and Local Government.

Department for Communities and Local Government (2016b) *Estate Regeneration National Strategy: Resident Engagement and Protection*. London: Department for Communities and Local Government.

Department of the Environment, Transport and the Regions (2000) *Planning Policy Guidance Note 3: Housing*. London: DETR.

European Convention on Human Rights, see *Convention for the Protection of Human Rights and Fundamental Freedoms*.

Future of London (2016) *Delivering Estate Renewal: A Briefing from London Housing Leaders*. London: Future of London.

Gray K and Gray S (2009) *Elements of Land Law*. Oxford: Oxford University Press.

GVA Bilfinger (2015) *Winstanley and York Road Estate Renewal: Planning Parameters Document*. London: London Borough of Wandsworth.

Hall P (1999) *Cities in Civilisation: Culture, Innovation and Urban Order*. London: Phoenix.

Hanna K, Oduwaiye A and Redman P (2016) *Another Storey: The Real Potential for Estate Densification*. London: Centre for London.

Harwood E (2015) *Space, Hope and Brutalism: English Architecture 1945–1975*. New York and London: Yale University Press, in association with Historic England.

Hill D (2014) Regenerating Southwark: Urban renewal prompts social cleansing fears, *Guardian*, 7 October. Accessed 19 January 2018 at www.theguardian.com/society/2014/oct/07/southwark-london-regeneration-urban-renewal-social-cleansing-fears.

Hill Z (2015) *Application for Confirmation of the London Borough of Barnet (West Hendon Regeneration Area) Compulsory Purchase Order No. 1 2014: CPO Report to the Secretary of State for Communities and Local Government*. London: The Planning Inspectorate.

HTA, Levitt Bernstein, Pollard Thomas Edwards and PRP (2016) *Altered Estates: How to Reconcile Competing Interests in Estate Regeneration*. London: Altered Estates.

The Labour Party (1945) *Let Us Face the Future: A Declaration of Labour Policy for the Consideration of the Nation*. London: National Executive Committee of the Labour Party.

Law Commission (2001) *Compulsory Purchase and Compensation: Disregarding 'The Scheme' – A Discussion Paper*. London: The Law Commission.

Law Commission (2003) *Towards a Compulsory Purchase Code: (1) Compensation – Final Report*. London: The Stationery Office.
Law Commission (2004) *Towards a Compulsory Purchase Code: (2) Procedure – Final Report*. London: The Stationery Office.
London Assembly Housing Committee (2015) *Knock it Down or Do It Up? The Challenge of Estate Regeneration*. London: Greater London Authority.
Loveland I (2018) *Constitutional Law, Administrative Law, and Human Rights: A Critical Introduction*. 8th ed. Oxford: Oxford University Press.
McLaughlin G (2015) *Estate Regeneration Sourcebook*. London: Urban Design London.
Mayor of London (2004) *The London Plan: Spatial Development Strategy for Greater London*. London: Greater London Authority.
Mayor of London (2011) *The London Plan: Spatial Development Strategy for Greater London*. London: Greater London Authority.
Mayor of London (2016) *Homes for Londoners: Draft Good Practice Guide to Estate Regeneration*. London: Greater London Authority.
Mayor of London and Southwark Council (2012) *Elephant and Castle: Supplementary Planning Document (SPD) and Opportunity Area Planning Framework (OAPF)*. London: Greater London Authority and London Borough of Southwark.
Office of the Deputy Prime Minister (2004) *ODPM Circular 06/2004: Compulsory Purchase and the Crichel Down Rules*. London: Office of the Deputy Prime Minister.
Pearce R (1994) *Attlee's Labour Governments 1945–51*. London: Routledge.
Prince's Foundation and The Hyde Group (2015) *Estate Regeneration: Six Steps to Success*. London: Prince's Foundation.
Roberts P and Sykes H (eds) (2000) *Urban Regeneration: A Handbook*. London: Sage Publications.
Royal Commission on the Distribution of the Industrial Population (1940) *Report* (M. Barlow, chairman). London: HMSO.
Secretary of State for Communities and Local Government (2015) *Re: The London Borough of Barnet (West Hendon Regeneration Area) Compulsory Purchase Order No. 1 2014*. London: Department for Communities and Local Government. Accessed 23 January 2018 at http://west-hendon.co.uk/planning/cpo1-public-inquiry/.
Secretary of State for Communities and Local Government (2016) *Re: The London Borough of Southwark (Aylesbury Estate Site 1B-1C) Compulsory Purchase Order 2014*. London: Department for Communities and Local Government.
Southwark Council (2007) *Heygate Action Plan – June 19 2007*. London: London Borough of Southwark.
Southwark Council (2010) *Aylesbury Area Action Plan*. London: London Borough of Southwark.
Tallon A (2010) *Urban Regeneration in the UK*. London: Routledge.
Van Kempen R, Dekker K, Hall S and Tosics I (2005) *Restructuring Large Housing Estates in Europe*. Bristol: The Policy Press.
Wandsworth Council (2014a) *Alton Area Masterplan: Council Tenant Rehousing Information Booklet*. London: London Borough of Wandsworth.
Wandsworth Council (2014b) *Alton Area Masterplan: Leaseholder and Freeholder Rehousing Information Booklet*. London: London Borough of Wandsworth.
Wandsworth Council (2015) *Winstanley and York Road Regeneration: Update Booklet*. London: London Borough of Wandsworth.
Whelan F (1980) Property as Artifice: Hume and Blackstone. In: Pennock J and Chapman J (eds) *Nomos XXII*: Property, 101–129. New York: New York University Press.

Cases

Pointe Gourde Quarrying and Transport Co Ltd v Sub-Intendent of Crown Lands [1947] AC 565 PC.

Waters and others v Welsh Development Agency [2004] UKHL 19.

Note on UK legislation. The original (as enacted) and revised versions of all UK legislation, both primary and secondary, can be accessed through *www.legislation.gov.uk*. This website is managed by The National Archives on behalf of Her Majesty's Stationery Office (HMSO).

5 Compulsory property acquisition for urban densification in Germany

Juliane Albrecht

Introduction

Many of Germany's cities are currently facing particularly high demand for housing due to high levels of immigration (in particular as a result of the influx of refugees), reurbanization and steadily declining household sizes. For these reasons, city planners are turning to urban densification. At the same time, the number of completed housing developments is in many places failing to meet the market demand for new dwellings. The creation of affordable housing is therefore at the top of the political agenda in Germany.

In July 2014, the Federal Ministry of Building and Urban Affairs together with the federal states, municipal associations, the housing and construction industry and other social actors set up the "Alliance for Affordable Housing and Construction" as a central instrument aimed at promoting housing construction. The aim of the alliance partners was to jointly improve the conditions for the construction and upgrading of good quality housing, preferably in the affordable market segment, and to expand the supply of dwellings in areas suffering from a housing shortage. A ten-point programme (the so-called "housing construction offensive") was derived from the recommendations of the alliance, which were adopted in November 2015. The programme encompasses a range of measures, including the aim of increasing the density of housing estates as well as developing brownfields and vacant plots.

A significant potential for residential construction is therefore seen in the activation of infill development. Since the Law to adapt the Federal Building Code to EU Directives of 24 June 2004 (EAG Bau, 2004) has entered into force, infill development has been anchored in the Federal Building Code (*Baugesetzbuch – BauGB*) as an urban planning model, subsequently strengthened by several BauGB amendments. For example, § 1a Para. 2 S. 1 Hs. 2 BauGB states that "in order to reduce the additional consumption of land for construction, the potentials of municipal developments through land restoration, densification as well as other measures to realize infill development should be exploited" (BauGB, 2004). The model of infill development is making an important contribution to the Federal Government's goal of limiting the consumption of land for settlement and transport purposes to less than 30 hectares per day by 2030 (see, most recently, Federal Government, 2016, 158ff.).

In 2006, infill development was strengthened by the introduction of so-called "development plans for infill development" in accordance with § 13a BauGB, which can be drawn up in an accelerated procedure (BauGB, 2006). The 2013 amendment to the *Baugesetzbuch* (BauGB, 2013) also anchored the priority of infill development into the planning guidelines of § 1 Para. 5 BauGB. Furthermore, § 1a Para. 2 S. 4 of the BauGB includes an obligation to submit reasons for the conversion of agricultural or forest land into building land. Such reasons must be based on investigations into the potentials of infill development, including in particular brownfield sites, vacant buildings, vacant plots and other sites suitable for urban densification (BauGB, 2013). Recently, § 6a of the Federal Land Utilization Ordinance (*Baunutzungsverordnung – BauNVO*) introduced the category of the *urban area*, which is intended to foster the development potential of inner city areas by means of efficient forms of construction with an appropriate mixture of residential and non-residential buildings ((BauNVO, 2017).

Clearly, the priority of infill development requires the consistent exploitation of existing infill development potentials. This can fail, in particular, because owners may be unwilling to sell plots, whether for personal or financial reasons (DV, 2016, 80). On the one hand, owners may wish to retain land in order to pass this on to their heirs, for example. On the other hand, land often remains unsold because plots are in the possession by a scattered group of part-owners or because the owners do not manage the plot for financial or personal reasons. Moreover, in view of the historically low long-term capital market interest rates, owners of potential building land do not see any economically viable investment alternatives. Increasingly, land is being held and acquired by private investors as a form of financial speculation (DifU and vhw, 2017, 16).

The question arises as to whether the owners of inner city plots who are unwilling to offer the land up for sale can be expropriated in order to activate these sites for the purpose of urban densification. The legal basis for this is specified in §§ 85 ff. BauGB, which regulates the instrument of expropriation for the purpose of urban planning. In the following sections the historical development and constitutional framework conditions of expropriation, the admissibility requirements, legal consequences and procedural regulations from the perspective of achieving urban densification are analysed. On this basis, the question is answered as to whether the expropriation of urban land is a suitable and sufficiently robust instrument to achieve higher levels of building density. In addition, an overview is given of current proposals to legally strengthen infill development as well as the role in this context of expropriation for urban development.

Historic and constitutional framework of expropriation for urban development purposes

The expropriation of land for urban development has a long tradition in Germany and is closely linked to the country's constitutional framework (see Albrecht, 2018, 294 ff.).

History of the expropriation of property for urban development purposes

Historically, the expropriation of property for urban development is rooted in nineteenth century laws promulgated by the federal states in the wake of the industrial revolution, aimed at removing from private ownership land required for public works such as rail or road construction (Runkel, 2009, preliminary remarks to 85 ff., no. 9). At the beginning of the twentieth century, the housing shortage that arose after World War I necessitated more standard and simplified expropriation regulations than found in the existing state laws. As a consequence, the ordinance on the rectification of the most urgent housing shortage of 9 December 1919 (*Behebungsverordnung*) was adopted by the German Reich, providing for a simplified acquisition of small and medium-sized flats.

After World War II, an urgent reform of existing urban planning provisions, especially the regulations on expropriation, was required to promote the reconstruction of destroyed towns and villages in order to accommodate millions of resettled persons and refugees. In some states the *Behebungsverordnung* was suspended, whereas other states questioned its validity and compatibility with the property right of Art. 14 Basic Law (Federal Government, 1951, 16). For similar reasons, the so-called Reconstruction Laws (*Aufbaugesetze*), which had been adopted by the states even before the foundation of the Federal Republic of Germany in 1949, were found to be problematic in terms of constitutional law. The *Aufbaugesetze* extended the possibilities of expropriation, including not only the procurement of building land but also other measures of urban development. Expropriation was regulated in these laws in different ways.

The Procurement of Building Land Act of 3 August 1953 (*Baulandbeschaffungsgesetz*) (BaulBG, 1953) was adopted to overcome this fragmentation of expropriation law at least for the purposes of housing construction. It was designed to provide a workable regulation which nevertheless conformed with the constitutional requirements for legal clarity and legal certainty. Due to the significance of expropriation measures for all parties involved, any element of uncertainty regarding the legal force of the decisions of the expropriation authority should a priori be excluded for constitutional reasons (Federal Government, 1951, 16–17). The *Baulandbeschaffungsgesetz* was primarily intended to supply land for housing by legislating for the compulsory sale of land to the state, which then sold such land to private individuals for the purpose of residential development. The legislative focus was to meet the urgent need for building land. However, consideration was also given to the later integration of the law into a comprehensive regulation of the entire corpus of building and land law (Federal Government, 1951, 17).

This comprehensive regulation was established by the Federal Building Act of 23 June 1960 (*Bundesbaugesetz*) (BBauG, 1960). In line with the general objectives of urban planning, it permitted municipal expropriation not only for the purpose of procuring building land but more generally in regard to building use in accordance with a municipally binding land use plan. Moreover, under certain conditions, expropriation of property could also be requested by private parties who wished to make use of a building. This extension of the purpose of expropriation aimed at enabling the implementation of urban land use planning (Federal Government, 1958, 86). Further expropriation purposes were supplemented by successive amendments of the Federal Building Code in 1976 and in 2004.

Constitutional basis for expropriation

Today, under Germany's constitution, the expropriation of property is closely regulated by guarantees of property ownership laid out in Art. 14 of the Basic Law (*Grundgesetz* – GG). The content of ownership is not stipulated by the constitution but is determined by the sub-constitutional legislator. This includes ownership according to civil law (§§ 903 ff. Civil Code, *Bürgerliches Gesetzbuch – BGB*) as well as all asset-related rights such as heritable buildings rights and the general right of pre-emption but *not* expectations of future profits and likely financial gain. The prevailing opinion is that protection extends to established and operating businesses insofar as their existence is threatened. The rights of local residents are only protected in core areas such as the permanent right of access, whereas additional regulations can be specified by the sub-constitutional legislator.

The guarantee of ownership distinguishes between expropriation (Art. 14 Para. 3 GG), on the one hand, and the contents and limits of ownership (*Inhalts- und Schrankenbestimmung*), on the other (Art. 14 Para. 1 Clause 2 GG). According to the jurisdiction of the Federal Constitutional Court (*Bundesverfassungsgericht, BVerfG*), expropriation is defined as the full or partial revocation of concrete, subjective legal positions (BVerfG, Ruling of 15.07.1981). In contrast, a determination on the contents and limits of property occurs when the legislator specifies rights and obligations in a general and abstract fashion without regard to individual persons or cases. Examples are limitations of use for plots for reasons of nature conservation or preservation of sites of historic interest. The *BVerfG* bases its distinction between expropriation, on the one hand, and the contents and limitations of ownership, on the other, not on the gravity of the intrusion but qualitatively.

According to Art. 14 Para. 3 S. 2 GG, expropriation can be achieved in two ways: through the law as legal expropriation (*Legalenteignung*) or pursuant to some law as administrative expropriation (*Administrativenteignung*). Legal expropriation empowers the legislator to revoke specific ownership rights directly

(Art. 14 Para. 3 S. 2 1. Alt. GG). In regard to administrative expropriation, on the other hand, the legislator empowers administrative bodies to revoke property ownership of individuals by means of administrative action (Art. 14 Para. 3 S. 2 2. Alt. GG). Expropriation for the purpose of urban development according to §§ 85 ff. BauGB is framed as administrative expropriation, i.e. the regulation of expropriation in the BauGB in the form of an administrative action.

In regulating expropriation, the legislator has to observe relevant constitutional requirements. Therefore, according to Art. 14 Para. 3 S. 1 GG, expropriation is only permissible for the general good. Furthermore, expropriation as well as all violations of property rights must obey the legal principle of proportionality, i.e. they must be suitable, necessary and proportional in the realization of the pursued aim. Finally, the law conferring the power of expropriation must also regulate the form and extent of compensation (the so-called *Junktim Clause*, Art. 14 Para. 3 S. 2 GG) (BVerfG, Ruling of 26.10.1977). Compensation is determined by an equitable balancing of the public interest as well as the interests of all affected parties (Art. 14 Para. 3 S. 3 GG).

Prerequisites of expropriation according to BauGB

Reflecting the constitutional empowerment of Art. 14 Para. 2 S. 2 GG, the legislator has framed expropriation for the purpose of urban land use planning within §§ 85 ff. BauGB as an instrument to ensure well-ordered urban development. The subject of expropriation can be the ownership rights to land or individual plots including the appurtenances of a plot, which can either be fully expropriated or encumbered with a right of use for the beneficiary of expropriation (§ 86 Para. 1 No. 1 BauGB). In addition, other rights to land can be removed or encumbered such as easements and heritable buildings rights (§ 86 Para. 1 No. 2 BauGB). The individual conditions upon which expropriation is contingent will be described in the following sections.

Permissible purposes of expropriation, § 85 BauGB

§ 85 Para. 1 BauGB specifies in numbers 1 to 7 the purposes which may be served by expropriation. It names in detail those conditions under which expropriation is admissible for the public good (cf. Art. 14 Para. 3 S. 1 GG). (Just, 2004, 392). The permissible purposes of planning are the following:

- expropriation to assist urban planning (No. 1);
- use of undeveloped plots or those developed to a very low level (No. 2);
- expropriation for land compensation (No. 3);
- expropriation to compensate revoked ownership rights (No. 4);
- enforcement of building orders (No. 5);
- preservation of a building structure subject to a preservation statute (No. 6);
- preservation or removal of building structures in urban redevelopment areas (No. 7).

The expropriation purposes according to No. 1, No. 2 and No. 5 are of particular importance for the expropriation of land or rights with the aim of urban densification. In the following paragraphs, it will be explained in more detail to what extent these can support the goal of urban densification.

Expropriation to assist urban planning

Expropriation to assist urban planning under § 85 Para. 1 No. 1 BauGB is the most important reason for the public expropriation of property. Its intention is to use a plot or to prepare a plot for use according to the designations in the municipal binding land use plan (*Bebauungsplan*). The permissible designations of such a land use plan are laid out in detail in § 9 BauGB (e.g. designations on the type and degree of building and land use). The purpose of expropriation as specified in § 85 Para. 1 No. 1 BauGB is to implement a municipal land use plan and its various designations. This is termed *planakzessorische Enteignung* (expropriation to assist urban planning) (Just, 2004, 392). The designations laid down by a municipality within a binding land use plan as permitted under § 9 BauGB can thus serve as valid reasons for expropriation (Dolde, 1991, 234). Since the binding land use plan determines not only the type but also the extent of construction, it can contribute to urban densification by determining a large building footprint or a high number of storeys.

Expropriation for the use of undeveloped plots or those developed to a very low level

The purpose of expropriation to make use of undeveloped plots within built-up areas designated by § 85 Para. 1 No. 2 BauGB may also be used for urban densification. Accordingly, undeveloped plots or those developed to a very low level which do not lie within the area covered by a binding land use plan but which lie within a built-up area can be expropriated for use or can be supplied for some use in accordance with regulations under building law (cf. § 34 BauGB). Expropriation for the purpose of infill development is explicitly mentioned in § 85 Para. 1 No. 2 BauGB. This constitutes a particular public good as it serves to reduce land consumption (Halama, 2009, § 85 no. 44). The development of vacant plots can also be enforced by issuing a building order and, if necessary, subsequent expropriation (BVerwG, Ruling of 15.2.1990).

Expropriation to enforce building orders

According to § 85 Para. 1 No. 5 BauGB, expropriation is also admissible in order to enforce a previously unsuccessful building order under § 176 BauGB by revoking ownership of a plot and offering this for sale to those willing to carry out the order (termed *Anschlussenteignung nach Baugeboten*) (expropriation to fulfil a building order). The expropriation can also be demanded by the municipality in favour of a party willing to develop the land (§ 87 Para. 3 S. 2

BauGB). First, it is necessary that attempts be made to enforce the building order by means of an administrative act, e.g. threat of a fine. If this repeatedly fails, then the municipal authorities can investigate whether the intended development can instead be achieved through expropriation (BVerwG, Ruling of 15.2.1990). In this case, however, additional benefits to the public good must be named to justify expropriation (Scheidler, 2016, 512). Since building orders are issued in order to develop vacant plots, they contribute to urban densification. However, in practice building orders are hardly ever issued (Stüer, 2009, 672), so the practical relevance of this form of expropriation purpose is low.

Other purposes of expropriation

Expropriation for land compensation and expropriation to compensate revoked ownership rights under § 85 Para. No. 3 and 4 BauGB are only indirectly relevant for urban densification, as they provide compensation in the form of land or grant ownership rights to resolve other cases of expropriation (Runkel, 2009, preliminary remarks to 85 ff., no. 8). Therefore, they are also called "assisting measures". The background to such *Ersatzlandenteignung* (§ 85 Para. 1 No. 3 BauGB) is the regulation in § 100 BauGB that if ownership meets certain conditions, the owner can apply for compensation against expropriation in the form of land. The background to *Ersatzrechteenteignung* (§ 85 Para. 1 No. 4 BauGB) is regulated by § 101 BauGB, whereby the expropriated owner may apply for compensation in the form of other rights.

The purposes of expropriation designated in § 85 Para. 1 No. 6 and 7 BauGB are less important for urban densification. § 85 Para. 1 No. 6 BauGB admits expropriation for the purpose of preserving a building structure situated within an area covered by a preservation order. In such cases, the plot can be expropriated and sold to a new owner who agrees to preserve the building structure. Clearly, this regulation does not contribute to urban densification but merely to preserving the status quo. According to § 85 Para. 1 No. 7 BauGB, expropriation can also serve the purpose of preserving or removing building structures in a legally designated urban redevelopment area under § 171d Para. 1 BauGB. Urban redevelopment areas are those afflicted by a considerable loss of essential urban functions (e.g. caused by high rates of long-term vacant housing). Here, however, the focus is rather on the removal of buildings than urban densification.

Requirements for the admissibility of expropriation, §§ 87 ff. BauGB

Reflecting the constitutional principle governing expropriation as laid out in Art. 14 Para. 3 S. 1 GG, the requirements for expropriation are formulated in §§ 87 ff. BauGB.

The general good

Corresponding with the requirements specified in Art. 14 Para. 3 S. 1 GG as well as the legal principle of proportionality, expropriation is only admissible in individual cases where this serves "the general good" and the purpose of expropriation cannot reasonably be achieved in any other way (Art. 87 Para. 1 BauGB). This regulation links the general good with the constitutional principle of necessity as applied to individual cases.

It insufficient to say that the form of land use specified in the municipal land use plan and named as the purpose of expropriation as designated in § 85 Para. 1 No. 1 BauGB necessarily serves the general good in the same way as every other planning measure. Rather, expropriation is only admissible if some *extraordinary* objective and concrete interest exists for the realization of a specific development measure, one that goes beyond the standard public interest in a planned development and which necessitates the violation of ownership rights for the general good. In other words, there must be some "particularly vital and urgent public interest" (BVerfG, Ruling of 24.3.1987). Use of the term "general good" implies a balancing of wider social benefits against the legal protection of ownership (Just, 2004, 395). In this way, expropriation is only admissible if it is found to be suitable, necessary and proportional in some significant fashion (Breuer, 2015, § 87 no. 10).

The stipulations of the land use plan, which comply with the *Abwägungsgebot* of § 1 Para. 7 BauGB (requiring the planning authority to weigh up all public and private interests), realize the general good at a *first level* and only *abstractly*. This is followed during the expropriation procedure by a further concretization of the general interest guided by the principle of necessary legal expropriation (cf. § 87 Para. 1 BauGB) and directed towards the individual case (Battis, 2016, § 87, no. 2). In some cases, there is even talk of a three stage concretization of the general good (Halama, 2009, § 87 no. 6; Runkel, 2010, § 85 no. 13 ff.; Breuer, 2015, § 85 no. 33):

- *First*, within the framework of the strict limits of § 85 Para. 1 S. 1 BauGB as to the designated specifications of a development plan in accordance with § 9 Para. 1 to 3 BauGB;
- *Second*, by the concretization of this range of potential specifications on the part of the municipality when drawing up the development plan; and
- *Third*, by the expropriation authority during the expropriation procedure when evaluating whether the planning specifications of the development plan are in the public interest as required by § 87 BauGB.

Referring to two decisions of the Federal Constitutional Court (BVerfG), it is sometimes demanded that expropriation for the purposes of realizing a particular project be expressly regulated by law, e.g. by the expropriation laws of the federal states, such as Art. 1 Expropriation Law of Bavaria (BayEG) and § 2 Expropriation Law of Saxony (SächsEntEG) (de Witt, 2017, 99–100). The decision of the BVerfG on *Bad Dürkheimer Gondelbahn* states that expropriation pursuant to Art. 14 Para. 3 S. 2 of the Basic Law is only permissible on the basis of a law that determines the concrete purpose of expropriation (BVerfG, Ruling of 10.3.1981). The *Boxberg Ruling* states that urban land use plans drawn up by the municipality are not suitable for making this decision due to the insufficient legal and democratic competence of municipal authorities (BVerfG, Ruling of 24.3.1987). Also § 87 Para. 1 BauGB does not contain any provision for achieving the goal of the general interest, instead only repeating the wording of Art. 14 Para. 3 S. 1 of the Basic Law (BVerfG, *ibidem*). Nevertheless, § 85 Para. 1 No. 1 BauGB was not declared unconstitutional by the BVerfG. The prevailing opinion, therefore, is to leave it to the expropriation authority to determine the necessity of expropriation.

Irrespective of this, however, expropriation appears to be hardly justified if it is only to achieve the general objective of urban densification. Instead, further reasons are required to justify the permanent withdrawal of ownership, reasons that must be strongly rooted in the general interest. This could be the case, for instance, if measures of urban densification also serve to improve the supply of essential public goods and services (e.g. the provision of energy or the establishment of schools and hospitals). Expropriation may then be justified also if they are not provided by local authorities but by some private party (BVerfG, Ruling of 18.2.1999; Halama, 2009, § 87 no. 59; Battis, 2006, 125). However, if the general good does not directly benefit from the business purpose but is rather an indirect consequence of business activities, then the Federal Constitutional Court views expropriation as inadmissible (BVerfG, Ruling of 24.3.1987).

In the light of this ruling, it seems highly unlikely that plots of land designated in the municipal land use plan as serving private purposes could be expropriated for the benefit of private developers if the intended development serves the general good only to the extent that it satisfies local demand for housing and business premises (Halama, 2009, § 87 no. 60). This is also true if the planning of this development is also aimed at urban densification. The case is rather different if, for example, the municipal demand for housing and workplaces is particularly urgent. Here too, however, carefully consideration must be paid to the owner's interests and to whether expropriation is actually necessary to meet the municipal demand. The same applies with regard to the expropriation purpose of § 85 Para. 1 No. 2 BauGB (use of undeveloped plots or those developed to a very low level). While the development of vacant plots is in the public interest, only in special cases does it seem justified to create by means of expropriation the preconditions for another private person to fulfil this task instead of the private landowner (Halama, 2009, § 87 no. 60).

In any case, expropriation is only admissible where its purpose cannot be reasonably achieved by any other means (§ 87 Para. 1 S. 1 2nd Clause BauGB, see also Federal Court of Justice – *Bundesgerichtshof* – BGH, Ruling of 25.10.2001). In other words, it is first necessary to investigate whether the purpose to be served by expropriation cannot be realized by another approach that violates the rights of the affected party to a much lesser degree (Battis,

2006, 125). Expropriation must always be the "last resort" (Just, 2004, 395). Clearly, this is not the case if, for example, similarly suitable public plots are available (BVerfG, Ruling of 19.12.2002). Expropriation may also be precluded if the purpose of expropriation can be achieved by other instruments less burdensome to the owner, such as reallocation (§§ 45 ff. BauGB) or the exercise of a pre-emption right (§§ 24 ff. BauGB).

Obligation to undertake purchase negotiations and evidence for realization of expropriation purpose

§ 87 Para. 2 S. 1 BauGB also regulates to protect ownership rights against unreasonable expropriation by requiring that the applicant has made a serious but vain attempt to privately acquire the land subject to expropriation on acceptable terms. Private purchase therefore takes precedence over expropriation (Halama, 2009, § 87 no. 63). Negotiations must be conducted with serious intent and with the aim of achieving a resolution acceptable to all parties (BVerwG, Ruling of 19.10.1966). The owner must be granted a reasonable period for deliberation (Halama, 2009, § 87 no. 67).

The applicant must provide evidence that the land will be used for the designated purpose within a suitable term (§ 87 Para. 2 S. 2 BauGB). In particular, the applicant must clearly announce the intended use of the land to be expropriated and must show sufficient own funding or creditworthiness (Halama, 2009, § 87 no. 82). This requirement is aimed at preventing the hoarding of building land or other forms of misappropriated land use (Breuer, 2015, § 87 no. 40).

Compensation for expropriation, §§ 93 ff. BauGB

According to § 93 Para. 1 No. 1 BauGB and corresponding to the requirements of Art. 14 Para. 3 GG, expropriation must be compensated (see Albrecht, 2018, 302 f.). This is intended to provide the affected party with a suitable indemnity for the loss incurred by the revocation of ownership rights (Just, 2004, 399). Compensation may be demanded by any person whose rights are adversely affected by expropriation, resulting in some financial detriment (§ 94 Para. 1 BauGB). The beneficiary of expropriation is legally obliged to make compensation (§ 94 Para. 2 BauGB).

According to § 93 Para. 2 BauGB, compensation is provided for rights forfeited as a result of expropriation (cf. § 95 BauGB, termed *Substanzentschädigung*), and second, for property loss of other kinds arising through expropriation (cf. § 96 BauGB, termed *Folgenentschädigung*). Examples of such financial detriment are a reduction in the market value of remaining plots or costs arising in connection with removal activities as a result of expropriation. Compensation is normally provided in the form of money (cf. § 99 BauGB). However, under certain conditions specified in §§ 100, 101 BauGB, the owner may apply for compensation in the form of land or the granting of other rights.

The amount of compensation is determined by considering an equitable balance of the interests of the general good and the affected parties (cf. Art. 14 Para. 3 GG). It should enable the affected parties to acquire an object of equal value. Other disadvantages that may arise (e.g. forgone profits) do not have to be fully compensated. In this regard, the law on expropriation specifically uses the term *compensation* ("Entschädigung") rather than *damages* ("Schadensersatz"). Under § 95 Para. 1 BauGB, the amount of compensation is assessed on the basis of the current market value of the plot to be expropriated or of any other subject of expropriation. The market value is set at the time when the expropriation authority adjudicates on an application for expropriation.

Under § 194 BauGB, the market value is defined as the price which would be obtained in ordinary transaction at the time when the assessment is made, taking into account the existing legal circumstances and the actual characteristics, general condition and location of the property or other object of assessment, without consideration being given to any extraordinary or personal circumstances. Independent committees of valuation experts are charged with determining property values (cf. §§ 192, 193 BauGB). These regularly compile purchasing price data for properties and use this data to determine standard ground values. The intention of the legislator in establishing the committees of valuation experts and their purchasing price data was to provide an instrument to ensure market transparency. The details of how to determine standard market values are regulated by the ordinance on guidelines to determine the market values of property (ImmoWertV, 2010).

Expropriation procedure

The expropriation procedure is regulated by the provisions laid down in §§ 104 to 122 BauGB (see Albrecht, 2018, 303 f.). These reflect the procedural requirements relating to ownership rights as laid out in Art. 14 GG (Battis, 2016, preliminary remarks to §§ 85 ff., no. 13). Application for an expropriation procedure to the expropriation authority is always the last means which the benefiting party can resort to when all other efforts to secure an agreement satisfactory to all parties have failed.

The preliminary expropriation procedure begins with the submission of an application for expropriation to the municipality within whose territory the land subject to expropriation is situated (§ 105 BauGB). The intention of this provision is to ensure that the special knowledge that a municipality possess of local circumstances can be introduced at this preliminary stage to the actual expropriation procedure. The municipality is required to present the application for expropriation with its comment to the expropriation authority (normally the higher administrative authority of a state) within one month (§ 105 S. 2 BauGB).

The expropriation authority is required to grant the property owner, the applicant and those authorities affected by the expropriation the opportunity to make representations (§ 107 Para. 1 S. 3 BauGB). In cases where ownership is to be withdrawn, the expropriation authority is required to obtain an opinion on property values from an independent committee of experts in order to determine the amount of compensation (§ 107 Para. 1 S. 4 BauGB). All questions relating to expropriation and compensation are discussed at the hearing. The expropriation authority must act as an intermediary between all parties involved in the expropriation procedure and strive to achieve some form of agreement (§ 110 Para. 1 BauGB).

If the parties involved reach *agreement*, the expropriation authority can certify the transition of ownership or an encumbrance on ownership as well as the amount of compensation in the same way as a notary public. Where the parties can only agree on the transition or the encumbrance of ownership but not on the amount of compensation, the expropriation authority can also certify this as *partial agreement* (§ 111 Para. 1 S. 1 BauGB). In this case, the expropriation authority continues with the procedure to determine the amount of compensation.

If the parties involved cannot reach agreement, the expropriation authority passes an *Enteignungsbeschluss* (expropriation resolution) (§ 112 Para. 1 BauGB). This resolution determines the amount of compensation and the interest to be paid on this. The expropriation authority bases its decisions on the land values calculated by the committee of experts. The resolution on expropriation is an administrative act that can be contested by the parties involved by lodging a motion for a court ruling to be conducted by the regional court's Chamber for Building-Land Matters (§§ 217 ff. BauGB).

Once the resolution on expropriation becomes incontestable, the beneficiary of expropriation must render or deposit the financial compensation. Thereafter, any of the parties involved can request that the expropriation authority execute the resolution on expropriation (*Ausführungsanordnung*) (§ 117 Para. 1 BauGB). This order of execution effects the legal changes determined in the resolution on expropriation (§ 117 Para. 5 BauGB). The order of execution includes the putting in possession of the expropriated land on the designated day (§ 117 Para. 6 BauGB). The expropriation authority conveys to the land registry the change of legal status (for example, transfer of ownership, notification of encumbrances) with the request that these alterations of rights be entered in the land register (§ 117 Para. 7 BauGB).

Practical implementation

Although expropriation is basically applicable for the purpose of urban densification, it is rarely applied in practice. This is due to the strict focus on the public good as well as the requirement that each case of expropriation be considered individually. Clearly, the legal guarantee of ownership rights presents a major hurdle for the admissibility of expropriation. The group of potential beneficiaries is also considerably restricted by the fact that expropriation in favour of private ownership is only admissible in the case of a particularly urgent and vital public interest. In addition, for reasons of municipal policy, there are often reservations about the use of such a "blunt" instrument (DifU and vhw, 2017, 16).

The "Alliance for Affordable Housing and Construction" has identified the need for a suitably strong instrument to ensure the supply of urban sites for development. This led to a proposal to introduce a so-called "infill development measure (IDM)" (DV, 2016, 82–83). The IDM is intended to mobilize the dispersed and heterogeneous potential of plots suitable for development in core urban areas and to encourage landowners themselves to undertake the building work in a timely fashion. In this way the IDM fosters the infill development of cities and municipalities and supports the previously mentioned sustainability goal of the Federal Government to reduce land consumption over the long term.

Within the framework of the IDM, a statute must first to be drawn up to determine the respective land use potential in a sub-area of a municipality. Such a statute may only be drawn up for reasons of the general good. This is the case, for example, if there is a considerable demand for housing and workplaces which can be met by infill development or if the IDM serves to utilize and maintain the full capacity of available technical and social infrastructure. In addition to plots in the possession of uncooperative owners which are suitable for development, the IDM is also intended to activate land for which the building right still has to be created by an IDM development plan (DifU and vhw, 2017, 16).

The timely development of urban plots in accordance with the development concept of the municipality should be secured by a building obligation with a stipulated deadline, determined by the above-mentioned IDM statute or IDM development plan. If the owner does not fulfil the building obligation before the stated deadline, the municipality is to submit an appropriate offer for purchase on the open market. If the owner rejects this, the possibility of expropriating the property in question exists as a last resort. In both cases, there is also the possibility of providing a settlement in the form of a compensatory plot rather than financial compensation (DifU and vhw, 2017, 16).

In general, the main objective of the IDM should not be the transitory acquisition of a plot by the municipality, whether through purchase or expropriation. Rather, the intention is to secure the plot's development by the property owner or by some third party to whom the plot can be sold. The need for an IDM as well as the effectiveness, manageability and legal backing of a corresponding regulatory proposal for a BauGB amendment are currently being examined within the framework of a so-called *Planspiel* (experimental planning game to test the impact of various amendments). The outcome of this Planspiel and the legislator's decision on the incorporation of the instrument into law are eagerly awaited.

Conclusions

Expropriation of property for urban development purposes has a long tradition in Germany. It is regulated in the German constitution and is concretized by the Federal Building Code. The latter names in detail those conditions under which expropriation is admissible for the public good. It specifies several expropriation purposes for urban densification, especially expropriation to assist urban planning, expropriation for the use of undeveloped plots or plots developed to a very low level, and expropriation to enforce building orders.

According to the Federal Building Code and corresponding to the requirements of the constitution it is obligatory that expropriation must be compensated, either by money, by land or by other rights in lieu. The compensation has to be provided by the beneficiary of the expropriation. The amount of compensation is assessed on the basis of the current market value of the plot to be expropriated. The market value is defined as the price which would be obtained in an ordinary transaction at the time when the assessment is made. It is determined in a transparent process under the involvement of evaluation experts.

The Federal Building Code regulates for a strict expropriation procedure to be carried out by specially constituted agencies that investigate all conditions, legal and otherwise, regarding an expropriation application, gather necessary information and initiate safety measures with the land registry. The obligation to negotiate is an integral part of the procedure. The due process is also guaranteed by the right to a hearing, the involvement of independent experts, the neutral position of the expropriation authority, as well as a comprehensive range of legal protections.

Hence, along with building orders, expropriation for urban development is basically applicable for the purpose of densification. However, as the above discussion has shown, the practical use of expropriation is subject to strict constitutional requirements, limiting it to specific individual cases and therefore usually entailing a great deal of effort and considerable legal risks. Therefore, the instrument is rarely used to activate building plots for the purpose of urban densification.

Although infill development has become a popular planning model for urban development in Germany and indeed is now anchored in the BauGB, green space is still being consumed for the purposes of new construction. This hinders the realization of urban densification. The proposed "infill development measure" (IDM) could overcome these obstacles. Whether it will be incorporated into the Federal Building Code and whether it actually will contribute to urban densification remains to be seen.

References

Albrecht J (2018) Expropriation for urban development purposes in Germany. In: Gerber J-D, Hartmann T, Hengstermann A (eds) *Instruments of Land Policy. Dealing with Scarcity of Land*, 294–306. New York: Routledge.

BauGB (2004) *Baugesetzbuch* in the version promulgated on 23 September 2004 (*Federal Law Gazette* I S. 2414), last amended on 20 July 2017 (*Federal Law Gazette* I, 2808).

BauGB (2006) Gesetz zur Erleichterung von Planungsvorhaben für die Innenentwicklung der Städte (Law to ease urban planning and encourage the infill development of cities and towns) of 21 December 2006 (*Federal Law Gazette* I, 3316).

BauGB (2013) Gesetz zur Stärkung der Innenentwicklung in den Städten und Gemeinden und weiteren Fortentwicklung des Städtebaurechts (Law to strengthen the infill development of cities, towns and municipalities as well as other revisions to town planning law) of 11 June 2013 (*Federal Law Gazette* I, 1548).

BauGB (2017) Gesetz zur Umsetzung der Richtlinie 2014/52/EU im Städtebaurecht und zur Stärkung des neuen Zusammenlebens in der Stadt (Law to transpose directive 2014/52/EU into town planning law and to strengthen the new coexistence in the city) from 4 May 2017 (*Federal Law Gazette* I, 1057).

BaulBG (1953) Baulandbeschaffungsgesetz of 3 August 1953 (*Federal Law Gazette* I, 720).

BauNVO (2017) Baunutzungsverordnung in the version promulgated on 23 January 1990 (*Federal Law Gazette* I, p. 132), amended by Law from 4 May 2017 (*Federal Law Gazette* I, 1057).

Battis U (2006) *Öffentliches Baurecht und Raumordnungsrecht*, 5th ed. Stuttgart: W. Kohlhammer.

Battis U (2016) Preliminary remarks to §§ 85 ff. In: Battis U, Krautzberger M and Löhr R-P (eds) *Baugesetzbuch. Kommentar*, 13th ed. Munich: C.H. Beck.

Battis U (2016) § 85 BauGB, *ibidem*.

Battis U (2016) § 87 BauGB, *ibidem*.

BBauG (1960) Bundesbaugesetz of 23 June 1960 (*Federal Law Gazette* I, 341).

BGH, Ruling of 25.10.2001 – III ZR 76/01.

Breuer R (2015) § 85 BauGB In: Schrödter W (ed.) *Baugesetzbuch*, 8th ed. Baden-Baden: Nomos, 1449–1471.

Breuer R (2015) § 87 BauGB, *ibidem*, 1477–1500.

BVerfG, Ruling of 26.10.1977 – 1 BvL 9/72.

BVerfG, Ruling of 10.3.1981 – 1 BvR 92/71, 1 BvR 96/71, *Bad Dürkheimer Gondelbahn Ruling*.

BVerfG, Ruling of 15.07.1981 – 1 BvL 77/78.

BVerfG, Ruling of 24.3.1987 – I BvR 1046/85, *Boxberg Ruling*.

BVerfG, Ruling of 18.2.1999 – 1 BvR 1367/88 et al.

BVerfG, Ruling of 19.12.2002–1 BvR 1402/01.

BVerwG, Ruling of 15.2.1990–4 C 45.87.

BVerwG, Ruling of 19.10.1966–4 C 57.65.

DifU and vhw – Deutsches Institut für Urbanistik and Bundesverband für Wohnen und Stadtentwicklung e.V. (2017) *Bodenpolitische Agenda 2020–2030*, Berlin.

DV – Deutscher Verband für Wohnungswesen, Städtebau und Raumordnung e.V. (ed.) (2016) *Mehr Bauland für bezahlbaren Wohnungsbau*, Berlin.

de Witt S (2017) Die Herausforderungen einer Enteignung nach § 85 Abs. 1 No. 1 BauGB, *Zeitschrift für Umwelt- und Planungsrecht*, 37(3), 97–100.

Dolde K-P (1991) Städtebauliche Enteignung und planerische Abwägung. In: Franßen E *et al.* (eds) *Bürger – Richter – Staat*, *Festschrift für Horst Sendler*, 225–239. Munich: C.H. Beck.

EAG Bau (2004) Gesetz zur Anpassung des Baugesetzbuchs an EU-Richtlinien vom 24. Juni 2004 (*Federal Law Gazette* I, 1359).

Federal Government (1951) Begründung zum Entwurf eines Gesetzes über die vorläufige Regelung der Bereitstellung von Bauland (Justification for the draft law on the provisional arrangement for the provision of building land) from 28 May 1951, *BT-Drs. (Bundestag printed paper)* 1/2281.

Federal Government (1958) Explanatory memorandum to the draft of the Federal Building Code of 16 April 1958, *BT-Drs. (Bundestag printed paper)* 3/336.

Federal Government (2016) Deutsche Nachhaltigkeitsstrategie, Frankfurt am Main.
Halama G (2009) § 85 BauGB. In: Driehaus H-J *et al.* (eds) *Berliner Kommentar zum Baugesetzbuch*, 14th update. Cologne: Carl Heymanns.
Halama G (2009) § 87 BauGB. In: Driehaus H-J *et al.* (eds) *Berliner Kommentar zum Baugesetzbuch*, 14th update. Cologne: Carl Heymanns.
ImmoWertV – Immobilienwertermittlungsverordnung of 19 May 2010 (*Federal Law Gazette* I, 639).
Just J-D (2004) Förmliche Enteignung. In: Hoppe W, Bönker C and Grotefels S, *Öffentliches Baurecht*, 3rd ed., 390–405. Munich: C.H. Beck.
Runkel P (2009) Preliminary remarks to §§ 85 ff. BauGB. In: Ernst W, Zinkahn W, Bielenberg W and Krautzberger M (eds) *Baugesetzbuch*, 93rd update. Munich: C.H. Beck.
Runkel P (2010) § 85 BauGB, *ibidem*, 96th update.
Scheidler A (2016) Die Voraussetzungen der städtebaulichen Enteignung, *Zeitschrift für Umwelt- und Planungsrecht*, 36(11+12), 510–516.
Stüer B (2009) *Handbuch des Bau- und Fachplanungsrechts*, 4th ed. Munich: C.H. Beck.

6 Accommodating densification and social sustainability in the inner city

Case study of Griffintown in Montreal

Sébastien Darchen and Claire Poitras

Introduction

This chapter analyses the densification process of a former industrial neighbourhood in the inner city of Montreal known as *Griffintown* within the South-West Borough (see Figure 6.1) and the role of compulsory acquisition in this process. This chapter builds on previous work questioning the sustainability of housing developments in the South-West borough which has been undergoing revitalisation since the late 1980s (Poitras, 2009). The chapter considers the planning documents promoting densification and the role of compulsory acquisition in this process for the Montreal context. Specifically, the chapter studies the role of the development industry in encouraging compulsory acquisition in the revitalisation process of a central area in Montreal.

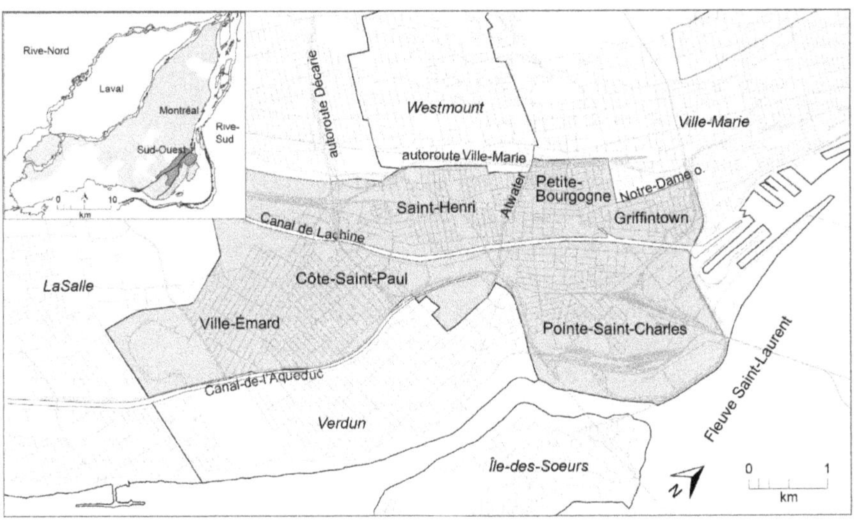

Figure 6.1 Location of Griffintown in the South-West Borough of Montreal.

Source: Institut National de la Recherche Scientifique, Urbanisation, Culture et Société, 2016.

This chapter specifically answers questions 2 and 3 of the book (Chapter 1). Furthermore, it examines if compulsory acquisition facilitates or not the implementation of social sustainability. We focus the analysis on urban design and heritage preservation in the process of densification. The chapter also examines the legislation aspects of compulsory acquisition in the Quebec context, and in doing so it also answers other questions asked in this book specifically for the Quebec context.

The chapter is divided into four main sections: the introduction, the methods, the analysis and the conclusion. In terms of the definition of social sustainability, we refer to Dempsey *et al.* (2011) as well as Chan and Lee (2008). Dempsey *et al.* (2011, 291) have identified that physical factors (e.g. urbanity, sustainable urban design, etc.) are contributory factors to the urban social sustainability of communities. Specifically, Chan and Lee (2008) have identified townscape design and the preservation of local characteristics as two components for sustainable urban renewal projects (among a total of six components) in the Hong Kong context. Ultimately, we are determining if the compulsory acquisition fostered social sustainability or not, considering the two components cited above that focus on the built form.

Brief overview of the case study

At the beginning of the nineteenth century, the area known as Griffintown[1] was a suburb inhabited by poor households. Between 1840 and 1870, the area benefited from the industrial activities happening in the South-West Borough of Montreal along the Lachine Canal (Shearmur, 2017). Considered as the industrial cradle of Canadian industry, the neighbourhood then became a mix of residential uses and small warehouses for manufacturing activities. Shearmur (2017, 212) calls it a working-class neighbourhood. Catholic Irish lived in the area, but the neighbourhood was also home to Francophone Catholic households and Anglophone Protestant families. Low-income working-class households were the majority but a middle class started to emerge as well in the second half of the nineteenth century (Barlow, 2009). Poitras (2017, 113) stated that Griffintown as a working-class neighbourhood was created in 1804. During the industrial era (from 1840 to 1940) the area now known as the South-West Borough was a place where a lot of manufacturing goods such as chemical products, food and clothing were produced. The area experienced an economic decline from the 1970s with large companies such as Northern Electric and the sugar factory Redpath Sugar leaving respectively in 1974 and 1976 (Poitras, 2017, 114). From the 1960s, the Griffintown area was zoned as an industrial area (more specifically for the storage of goods). This induced a loss of residents. The revitalisation of the South-West Borough through real estate development started from 2008. Overall the South-West Borough is experiencing rapid urban redevelopment that includes an intake of new households as well as new economic activities settling in the area: video game industry, urban design and architecture-related economic activities and cultural industries.

However, in 2011, despite fast redevelopment taking place, the population of the South-West Borough remained less than 40 years ago (Poitras, 2017, 114). The main challenge for the South-West Borough – and this includes Griffintown – is actually to "reinvent itself" while preserving its industrial identity (Poitras, 2017, 121).

Between 1970 and 1990 the area became a former industrial neighbourhood ripe for redevelopment (Shearmur, 2017). According to Shearmur (2017, 214), the area has experienced two types of urban renaissance. The first one is a revitalisation process driven by an Engineering School, l'École de Technologie Supérieure (ETS), which acquired land there and initiated the Innovation District project. The second urban renaissance is a real estate revitalisation process driven by private developers. The densification process driven by developers, and the role of compulsory acquisition in that process, is the focus of this chapter.

Methods

The methods of data collection included semi-structured interviews with selected stakeholders involved in the densification process. We have also performed a review of press articles between 1995 and 2017 to better understand the history of compulsory acquisition in the area. This period has been selected because it corresponds to the real estate boom that has changed the South-West Borough's cityscape. We also analysed official documents (Minutes of Municipal Council[2]) between 1990 and 2016 in relation to compulsory acquisitions of land or housing for the perimeter of Griffintown. The planning documents analysed were the Plan d'Urbanisme[3] (2004) which covers the whole City of Montreal and the two successive planning schemes specific to the Griffintown area: Plan Particulier d'Urbanisme[4] (PPU) (2008), PPU (2013).

Analysis

Which planning tool is driving the densification process in Griffintown?

In the first PPU issued in 2008 for the Peel-Wellington area, the approach was not very clear and the "project" was presented as a large-scale real estate project driven by one developer: Devimco. In the second PPU issued in 2013, the vision was articulated more around the principles of innovation and the preservation of the industrial historic built environment (Ville de Montréal, 2013, 16). The approach since 2013 is closer to an approach to redevelopment known as "urban regeneration". Urban regeneration can be defined as follows: A comprehensive and integrated vision and action which leads to the resolution of urban problems, and which seeks to bring about lasting improvement in the economic, physical, social and environmental condition of an area that has been subject to change (Roberts, 2000, 17).

This section analyses which planning tools were applied for the densification of the area. As stated by Breux and Bédard (2014), some large-scale redevelopment initiatives in Montreal in the early 2000s such the new Casino project and the Griffintown densification were large enough to be considered as an "Urban Project".[5] This concept is widely used in France to define large real estate redevelopments. Breux and Bédard (2014, 69) refer to Bailleul (2008) to define what an urban project is: "the urban project context facilitates the development of a coalition of stakeholders that will decide the project's objectives and discuss throughout the project while the objectives are being completed". However, the larger this coalition is, the higher the risk of conflict. Breux and Bédard (2014) labelled Griffintown as "un projet multi-fonctionnel d'envergure pan-Montréalaise".[6] But public consultations for the first PPU have been managed at the Borough level and not through the Office de Consultation Publique as should have been the case (Breux and Bédard, 2014, 77).

It is important to note that the name Griffintown has been promoted through new development projects led by developers, as part of this area was known as the district of La Petite-Bourgogne (Guillemard, 2015). According to Guillemard (2015), strategies related to economic development and revitalisation are driving the process rather than a logic of cultural development based on the preservation of existing industrial buildings in the area. The manufacturing industries and residents left the area earlier than in the case of the Angus Shops (the site of a former extensive railcar manufacturing and repairing facility) for example. The decline of Griffintown has been fast. It started with a change of zoning in 1966–1967 that prevented more residential development, the decline of industrial activities and the closure of the Lachine Canal in 1970 (Office de Consultation Publique de Montréal, 2012, 3).

The history of this area is very much connected to the Canal Lachine's history. After the Canal was closed to shipping in 1970, the economic decline of the area was rapid (Poitras, 2017, 108). The area stayed as it was and no economic revitalisation strategy was implemented for a while. This meant that some warehouses and industrial buildings had been abandoned for many decades when the time for revitalisation arrived. According to Guillemard (2015, 107), this increased the renovation costs for industrial buildings in Griffintown. The idea of revitalising the area started around the 1980s. Adjacent areas also went through major revitalisation initiatives, the Multimedia City being the main project. In November 2004, the area Peel-Wellington was marked as an area for "planification détaillée"[7] in the Plan d'Urbanisme[8] for Montreal. The City of Montreal approved the PPU in 2008. The developer Devimco leading the densification is known for the development of a lifestyle centre in the city of Brossard in Montreal's South Shore suburb and has very limited experience in leading densification in an inner city context (Breux and Bédard, 2014, 77). Devimco's real estate development was considered as a mega-project prioritising residential development but the concept was not very innovative: the area was to become more mixed use (residential, commercial activities, offices, entertainment) and a light rail was supposed to go through the area (Breux and Bédard,

2014). The Master Plan has been criticised as the new layout did not match at all the layout of streets that was typical of this industrial area. In other words, the developer presented a Master Plan that considered the area as a "friche industrielle"[9] (industrial wasteland) but the area had at least 55 buildings recognised as having architectural and historical value (Office de Consultation Publique de Montréal, 2012).

In terms of economic development related to the Innovation District led by ETS, McGill University and officially launched in 2003, Shearmur (2017, 222) calls it an "imaginary district" as it does not relate well with the history of the area. Shearmur (2017) also explains that the proximity of different firms in the New Economy sectors will not necessarily lead to the development of an innovative district. Shearmur (2017) observes that the discourse on innovation was used to justify decision-making in terms of urban planning and development for this area. But this discourse, according to Shearmur (2017), is disconnected from the history of the area.

The redevelopment scheme proposed by Devimco had an emphasis on public transport with a tramway line but also private transport and a mix of uses but the overall concept lacked innovation (Breux and Bédard, 2014; Shearmur, 2017). An open letter was published in the newspaper *Le Devoir* on 26 April 2008. It was written by architects from Montréal. The main point was that the Devimco project was a missed opportunity. The letter stated that the land available in Griffintown was an opportunity to apply innovative and architectural principles to this central and strategic area of Montréal. Referring to the BoO1 project in Malmo (Sweden) – considered as a model of sustainable development in Europe – the letter was asking for the City of Montreal to take control. The latest local plan (or PPU) issued in 2013 was an attempt to integrate those principles but according to Shearmur (2017, 217), between 2008 and 2013, many high-rise condominiums dramatically changed the urban landscape in Griffintown. Overall, the revitalisation process happening under the first PPU published in 2008 has been criticised for its lack of social sustainability, specifically on the issue of heritage. For instance, David Hanna (2012), Professor at UQAM, regrets that Montreal is not following other North American strategies towards heritage-based policy for urban revitalisation. Specifically, he is referring to Portland (Oregon, US), a model city in terms of sustainable development, which adopted a development plan that fosters high-tech industries while preserving cultural heritage (City of Portland, 2017).

As specified in Table 6.1 the area studied has densified under different planning schemes. The new PPU issued in 2013 is closer to the preferred principles of urban regeneration. The Bassins du Nouveau Havre project led by the Canada Land Company (CLC) will develop 2,000 dwellings including affordable housing and housing designed for families (Ville de Montréal, 2013, 10). The Master Plan was designed by a Montreal-based architecture firm Cardinal Hardy: the density is lower compared to the rest of the Griffintown perimeter. The fact that the land is owned by CLC facilitated the integration of sustainability principles (Interview 3). The latest PPU was based on public

Table 6.1 Timeline – Griffintown densification process

Date	*Main planning document*	*Main facts*
2004	Plan d'Urbanisme	Three objectives are specified for Griffintown: To preserve the industrial identity of the area; To densify and increase the mix of uses between economic activities and residential; Use the Peel basins (basins formerly used to load flour and grains of wheat for shipping) to increase the potential of the area for tourism.
2008	PPU Peel-Wellington	Plan is designed to serve the objectives of one developer, Devimco; Compulsory acquisition planned by the City of Montreal; The "project" is a real estate project.
2008–2009		Financial crisis – Architects and academics express discontent with Devimco's project.
2013	PPU Griffintown	Densification process is more connected to regeneration principles with an emphasis on innovation and preservation of industrial heritage; Bassins du Nouveau Havre project led by the Canadian Land Corporation (CLC) to be an example of sustainable densification.

Source: authors' research.

involvement to create an urban environment in accordance with what people want. The City of Montreal will have an accompanying role in the densification process. The City is likely to use compulsory acquisition to acquire land for the design of public spaces and to achieve the objectives stated in the most recent PPU (Interview 1).

The role of compulsory acquisition in the densification process of Griffintown

It is rather unusual for local governments in Quebec to use compulsory acquisition for a private developer; however, according to the Law on Planning and Urbanism[10] a local government such as the City of Montreal can use its right to acquire land or buildings in the context of a PPU when the objective is to help the developer to amalgamate the land inside the perimeter defined in the PPU. The developer must demonstrate that it owns or has obtained acceptance to sell from the landowners for a minimum of two-thirds of the area (Legis Quebec, 2017). However, the City of Montreal in the case of Griffintown was willing to reduce compulsory acquisition as much as it could; the strategy is then to give time to the developers to negotiate with the land owners about a price for the acquisition of land (Corriveau, 2008). The total cost associated with the compulsory acquisition of land (it includes the costs associated with the cleaning of contaminated soils evaluated at C$4.8 million) was C$65 million in 2008 for the acquisition of 25 land parcels in the whole perimeter for the first PPU Peel-Wellington (Corriveau, 2008). According to our research, the case of Griffintown for the first PPU approved in 2008 is an example of planned compulsory acquisition but most of the acquisition did not happen. There are two main reasons for this: first, the financial crisis; second, the fact that the developer was unable to deliver proof of finance in due time.

For example, on 28 April 2008, there was mention in the official documents of the compulsory acquisition of 21 land parcels in order to achieve the street layout stated in the first PPU (Resolution number CM08 0281–6); but then in a later decision on 23 February 2009, those compulsory acquisitions were cancelled (Official document number CM09 0085). When there have been compulsory acquisitions, there has been financial compensation. As an example, the City of Montreal agreed on 22 January 2001 with Minel Kupferberg (the land owner) to pay C$2,790,000 for the acquisition of two land parcels located between King and Queen Streets north of de la Commune Street (Resolution number 01 00160). The process of land acquisition is monitored by the Tribunal Administratif du Québec (TAQ). As a first step, there is negotiation between the City of Montreal and the landowner. If an agreement cannot be reached, TAQ will decide on the price by taking into account the value of the building, the loss of rent revenue, etc. (Affaires Municipales et Occupation du Territoire (Québec), 2017). The owner can also appeal to the Superior Court (Cour Supérieure) to oppose the land acquisition (TAQ, 2017). TAQ is the institution deciding on the price for the land acquisition but the Superior Court is the institution that adjudicates if the land acquisition process is appealed.

Our research reveals that the process for the Griffintown project is unusual. Basically for the first PPU, the developer (Devimco) came up with a plan for the area and the City of Montreal acquired land to bring about this plan. In 2008, the City of Montreal issued "Un avis de reserve foncière"[11] which put all land on hold for the project within the perimeter of the PPU. The following excerpt illustrates this point: "The City of Montréal has issued a reservation of land for two years; it can be renewed for another two years; it is not a common process but the city is allowed to do so within a PPU" (Lévesque, 2008). The PPU also was designed the suit the developer's plan with no interference from the Office de Consultation Publique (which conducts the usual public hearings process for large-scale initiatives in Montreal). In summary, we can say that the densification process during the first PPU is an example of planned compulsory acquisition that did not happen because of the financial crisis of 2008–2009 and of the financial difficulties experienced by Devimco.

The amounts for compensation are voted by the local council; it is a case by case matter and there is no information available on how the amount is calculated. The City acquired some land for the provision of public spaces and for the redesign of street layout. But as we mentioned most of the land acquisitions that were planned by the City were cancelled, meaning that the developer has been able in some cases to negotiate directly with the land owners. Our review of press articles reveals that the financial crisis in 2008 forced the main developer Devimco to revise the initial project. The City of Montreal thus cancelled the planned land acquisition that would have been allowed under the first PPU. Clearly, the review of the press articles as well as our interviews reveal that the financial crisis and the financial problems of the developer (Devimco) encouraged the City of Montreal to cease land holding (through compulsory acquisition) as Devimco's project would have been more modest given the real estate situation at the time (Corriveau, 2009).

In regard to the acquisition of housing, there were no attempts in the Griffintown perimeter, where the neighbourhood's housing was located. The only case of acquisition of housing close to the Griffintown perimeter was the building located at 780 St Remy in St-Henry which was demolished in 2015. The Ministry of Quebec Transport (MTQ) agreed to finance 50 per cent of a new housing project of 103 dwellings (at a total cost of C$24.5 million) to be located in Griffintown and completed in 2019 (Gaior, 2016). The first letters were sent to the tenants by the City of Montreal in 2009 and the land for the new housing project was purchased in 2014. The demolition of 780 St Remy was necessary due to the upgrading of the intersection of two major highways (Gaior, 2016). The review of press articles reveals that some residents have accepted monetary compensation and some have decided to be relocated in the housing projects proposed by the MTQ. This is an unusual process as there is no obligation to provide alternative housing in lieu but considering the residents' protests were well documented in the media, the MTQ felt it had to propose an alternative in terms of housing provision. Our research showed that the scale of compulsory acquisition was in the end not as large as expected.

Does compulsory acquisition necessarily promote social sustainability?

In this section, we analyse the final outcome in regard to social sustainability by focusing on two components: townscape design and the preservation of local characteristics identified by Chan and Lee (2008). Our aim is to answer question 2 of the book: *What specific public benefits are given for compulsory acquisition?* An analysis of a few case studies within the Griffintown perimeter shows that the discourse on public benefits is not clearly articulated. We also show that initiatives improving the social sustainability of the area happened outside the compulsory acquisition process. The fact that compulsory acquisition was cut back led to an improvement in social sustainability by allowing building owners to rehabilitate existing buildings like the New City Gas and the Darling Foundry. Figure 6.2 shows the two adaptive re-use initiatives, as well as the Rodier building built in 1875 that was supposed to be demolished for the electrical train project (light rail) promoted by the Caisse des Dépôts et Consignations du Québec[12] (CDPQ) but which was finally preserved along with the New

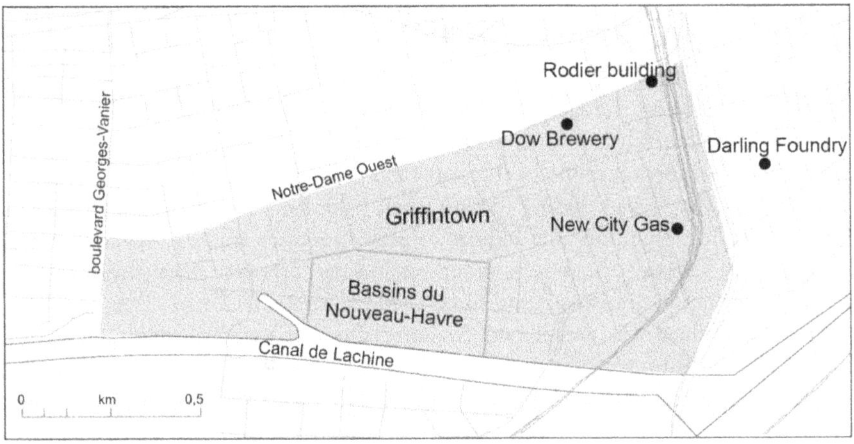

Figure 6.2 The Griffintown perimeter PPU (2013).

Source: Institut National de la Recherche Scientifique, Urbanisation, Culture et Société, 2016.

City Gas Building (Sioui, 2016). Figure 6.2 also includes the Dow Brewery building which was acquired to be part of the École de Technologie Supérieure (ETS) as part of the "Quartier de l'Innovation" project within Griffintown (Gass, 2017). Finally, the Bassins du Nouveau-Havre project is another attempt to serve the sustainability of the area led by CLC but outside the compulsory acquisition process described in the last section.

The densification of the Griffintown perimeter is still on-going but based on observation and our semi-structured interviews, we can already assess the outcomes of the process that started with the first PPU (2008) and the Devimco project. We focus on two case studies where the owners have been able to maintain ownership and change the use of the building to serve the social sustainability of the area.

The Darling Foundry is an example of a sustainable adaptive re-use initiative within Griffintown driven by individuals. The Darling Foundry occupies 3,500 m^2 in two adjoining former industrial buildings in Old Montréal. It presents exhibitions, makes studios available to local artists, and hosts international residencies, offering artists and the public fulfilling experiences in an inspiring context and with a sense of coherence that facilitates the comprehension of contemporary art (Fonderie Darling, 2017). The renovation of the building was funded by the provincial and federal governments as well as private partners (C$4 million in total was raised by "Quartier Ephémère", an association founded in 1994) and started in 2002. The Darling Brothers business was owned by the City of Montreal. At the time, the City of Montreal was open to new concepts to promote the dissemination of art in the area. The concept was initially from Paris (France) where the "hôpital éphèmere" project was developed. The Minister of Culture from Quebec liked the idea and decided to apply it in Montréal. In the early 2000s the association saved the building from demolition, and the City designated the building as a heritage building in 2001 (Interview 4). This example shows that negotiations around the compulsory acquisition between the City of Montreal and the owner have happened. The mayor at the time, Mr Bourque, believed that this adaptive re-use project would serve the social sustainability of the area (Interview 5).

The New City Gas building reconversion is an example of planned compulsory acquisition that did not happen. Public hearings in 2010 forced Mayor Tremblay to redirect the bus corridor around the property after the threat of compulsory acquisition was going to be used (Hanna, 2012). In 2016, the New City Gas building was threatened by compulsory acquisition again, this time because of a 67 kilometre light-rail transit system connecting downtown and the Trudeau Airport (Riga, 2016). New City Gas now hosts cultural activities including art galleries and places of artistic creation, as well as a well-run nightclub (Figure 6.3). The latter would be ranked as the most popular in Canada, receiving close to 5,000 night clubbers (hosting international superstar DJs such as Moby). This place of entertainment has repeatedly raised the issue of cohabitation with the inhabitants of the neighbourhood. Both the New City Gas

Figure 6.3 Old meets new. A view of nineteenth-century housing and contemporary condominium development.

Source: Denise Caron, 2017.

building built in 1859 and the Rodier Building built in 1875 have been threatened with compulsory acquisition by the light-rail transit system project but they were finally preserved through negotiations (Riga, 2016). The New City Gas is listed in the Provincial Government's directory of heritage buildings. This case highlights the need to use compulsory acquisition powers carefully. Our research has highlighted that compulsory acquisition strategies in the Griffintown case study did not serve the purpose of social sustainability but rather the interests of the Devimco developer. Socially sustainable projects such as New City Gas or the Darling Foundry are the results of negotiations with the City and happened outside the compulsory acquisition process.

On preserving the identity of the area, compulsory acquisition has had no effect at all. The Griffintown case study shows that compulsory acquisition was used to improve the street layout or to serve public transport projects. This would attract more developers in the area and ultimately change the urban landscape by densifying the residential built environment. Although history has been enhanced through the preservation of a few heritage buildings, notably in connection with the industrial past of the neighbourhood like the Darling Foundry and the New City Gas building, most of the new buildings in the Griffintown area have turned their backs on history. This is particularly noticeable with respect to residential types (see Figure 6.3). Thus, in recent years, a significant number of new residential towers have been built, the height of which exceeds that of the traditional built-up area. In Canada, the term Vancouvarism was coined to describe the development approach to addressing urban sprawl by intensifying the residential built environment. Established in the Greater Vancouver area, this urban design model, combining residential compactness, public transit use and access to quality green spaces, aims to promote a sustainable urban lifestyle (Brunet-Jailly, 2008). In Montreal, municipal officials and real estate developers have not necessarily sought to replicate the trendy residential model on Canada's West coast. Nevertheless Griffintown's residential densification has changed the shape of the urban fabric and the scales of the built environment. From now on, the rare red brick houses built in the nineteenth century co-exist with the very fenestrated residential towers. The condominiums developed on former vacant lots left empty by the departure of large manufacturing companies offer their occupants high views. Figure 6.3 highlights the challenge of promoting a densification project that is aesthetically appealing.

Conclusion

The City of Montreal justified compulsory acquisition in practice by advocating that Devimco's project would serve Montreal as a whole. Specifically, the new light rail associated with the first PPU was used as a component to enhance the sustainability of the overall project but research has shown that the innovation in terms of planning was limited for the first PPU (Shearmur, 2017). In the case of the first PPU, the development industry played a key role in promoting compulsory acquisition. The City of Montreal backed up the initiative of the developer prior to the financial crisis (2008–2009) as the City wanted regeneration to happen. As other large projects such as the Casino did not happen, the City of Montreal was willing to promote redevelopment in this central area of Montreal to raise land taxes.

As we have shown, the compulsory acquisition set up by the City of Montreal was planned to suit Devimco's project. It was a rather unusual process. In terms of public benefits, the City of Montreal claimed that the project would generate investments and foster economic revitalisation. The development industry has been very active in promoting compulsory acquisition to make sure the ambitious project – planned in the first PPU – would happen. As stated above, compensation is managed by TAQ, and there is no real specification on how the compensation is calculated. This research also shows that significant buildings such as New City Gas have been preserved through the initiatives of individuals sympathetic to heritage and arts and at some point the City had to give up the expropriation reserve on this property (Hanna, 2012). Overall, in the Montreal case, it has been unusual for the City to acquire land on behalf of a developer, so the Griffintown case study in its first phase (first PPU) is an exception in the Montreal context.

This chapter has answered questions 2 and 3. On question 2: *What specific public benefits are given for compulsory acquisition?* we can say that the public benefits discourse was organised around the ideas of economic innovation (Innovation District) and sustainability (public transport projects). The City of Montreal used those objectives to justify a rather unusual compulsory acquisition in the Montreal context. On question 3: *What role does the development industry play in facilitating, encouraging or promoting compulsory acquisition?* we have shown that the large compulsory acquisition process associated with the first PPU (Devimco's project) was promoted by the developer itself but due to both critics and the financial crisis, the compulsory acquisition process had to be scaled down by the City. Finally, this chapter has shown that social sustainability for the area has been promoted through initiatives (e.g. New City Gas building and the Darling Foundry) happening outside the compulsory acquisition process.

Notes

1 The name Griffintown comes from Mary Griffin who acquired several land parcels in the area around 1800 (Barlow, 2009).
2 For urban projects under a PPU scheme, the decisions are made at the City level and not at the Borough level.
3 Plan d'Urbanisme is a document presenting the main planning principles for the City of Montreal. It identifies areas – such as Griffintown – requiring special intervention embodied in a "Plan Particulier d'Urbanisme" (PPU).
4 A "Plan Particulier d'Urbanisme" (PPU) enables local governments to apply specific planning regulations for a selected area. It can include specific density requirements, mixed uses, etc.
5 The French word is "Projet Urbain".

6 Can be translated as "multi-functional urban project".
7 Can be translated as "area marked for detailed planning".
8 Can be translated as "City Plan".
9 Can be translated as "disaffected industrial area".
10 The law is: La loi sur l'aménagement et l'urbanisme.
11 Can be translated as "reservation of land".
12 Established in 1965, CDPQ manages the funds of large Québec-based public and para-public pension and insurance plans.

References

Affaires Municipales et Occupation du Territoire (Québec) (2017) Guide La prise de décision en urbanisme – Expropriation. Accessed 21 August 2017 at www.mamot.gouv.qc.ca/amenagement-du-territoire/guide-la-prise-de-decision-en-urbanisme/financement-et-maitrise-fonciere/expropriation/.

Bailleul H (2008) Les nouvelles formes de la communication autour des projets urbains: modalités, impacts, enjeux pour un débat participatif, *Métropoles*, 3, 98–139.

Barlow J (2009) *The House of the Irish: Irishness, History and Memory in Griffintown, Montreal, 1868–2009*. Doctoral Thesis, Montreal, Concordia University.

Breux S and Bédard M (2014) Lecture géopolitique d'un projet en devenir: Griffintown. In: Raynaud M, Lewis P and Hubert M (eds) *Les Grands Projets Urbains – Territoires, Acteurs et Stratégies*. Montréal: Presses de l'Université de Montréal.

Brunet-Jailly E (2008) Vancouver: the sustainable city, *Journal of Urban Affairs*, 30(2), 375–388.

Chan E and Lee G K L (2008) Critical factors for improving social sustainability of urban renewal projects. *Social Indicators Research*, 85(2), 243–256.

City of Portland (2017) Economic Development Planning. Accessed 11 July 2017 at www.portlandoregon.gov/bps/article/401339.

Corriveau J (2008) Le promoteur de Griffintown n'a pas fourni la garantie attendue, *Le Devoir*, 29 May.

Corriveau J (2009) Un autre clou dans le cercueil du projet Devimco à Griffintown, *Le Devoir*, 18 February.

Dempsey N, Bramley G, Power S, and Brown C (2011) The social dimension of sustainable development: defining urban social sustainability, *Sustainable Development*, 19(5), 289–300.

Fonderie Darling (2017) Mandate. Accessed 11 August 2017 at http://fonderiedarling.org/en/mandate.html.

Gaior C (2016) Exproprié à cause des travaux de l'échangeur Turcot: six ans pour être relocalisé, *Le Journal de Montréal*, 1 March.

Gass H (2017) "Innovation District" latest attempt at Griffintown development, *Spacing*. 16 July. Accessed 15 January 2017 at http://spacing.ca/montreal/2012/07/16/innovation-district-latest-attempt-at-griffintown-development/.

Guillemard A (2015) *La Place du Tourisme dans les Reconversions d'Anciens Territoires Industriels Montréalais: Comparaison entre Griffintown et les Shops Angus*. PhD Thesis, Montreal, University of Montreal.

Hanna D (2012) At risk in Griffintown: important pieces of Montreal's heritage, *Montreal Gazette*, 13 April.

Legis Quebec (2017) Loi sur l'aménagement et l'urbanisme. Accessed 10 July 2017 at http://legisquebec.gouv.qc.ca/fr/ShowDoc/cs/A-19.1.

Lévesque K (2008) Autre coup de pouce à Griffintown, *Le Devoir*, 17 January.

Office de Consultation Publique de Montreal (2012) Secteur de planification détaillée de Griffintown. Rapport de consultation publique, 13 Avril.

Poitras C (2009) Designing sustainability for whom? Recent housing developments in Southwest Montréal, *Local Environment*, 14(6), 515–528.

Poitras C (2017) L'axe du Canal Lachine et les quartiers du Sud-Ouest. Grandeur et misère du berceau de l'industrialisation du pays? In: Klein J-L and Shearmur R (eds) *Montréal la cité des cités*, 107–123. Montréal: Presses de l'Université du Québec.

Riga A (2016) Griffintown heritage buildings may be expropriated as part of light-rail plan, *Montréal Gazette*, 31 May.

Roberts P (2000) The evolution, definition and purpose of urban regeneration. In: Roberts P and Sykes H (eds) *Urban Regeneration. A Handbook*, 9–36. London: Routledge.

Shearmur R (2017) Le QI de Montréal – Quartier de l'innovation ou quartier imaginaire. In: Klein J-L and Shearmur R (eds) *Montréal la cité des cités*, 107–123. Montréal: Presses de l'Université du Québec.

Sioui, M (2016) Caisse des Dépôts. Le train nuira à des espèces menacés, *Le Devoir*. 29 July.

Tribunal Administratif du Québec (2017) L'expropriation. Vous avez des droits. Accessed 21 August 2017 at www.taq.gouv.qc.ca/documents/file/Depliant%20expro %20FRANCAIS-final%20(octobre%202005).pdf.

Ville de Montréal (2013) *PPU*. Montréal: Ville de Montréal.

Semi-structured interviews

September–October 2016 (Montreal)

1 Town Planner – South-West Borough, Montreal, 26 September.
2 Business owner in Griffintown, September.
3 Project manager, Canadian Land Company (CLC) Bassins du Nouveau Havre, 28 September.
4 Project manager, Bassins du Nouveau Havre, South-West Borough, Montreal, 14 October.
5 Manager, Fonderie Darling, 14 October.
6 Architect – Master Planning for Bassins du Nouveau Havre, 21 October.

7 The use of compulsory property acquisition and land readjustment in urban densification in Spain

Demetrio Muñoz Gielen and Marta Lora-Tamayo Vallvé

Introduction

In Spain, property needed for public infrastructure (roads, green infrastructure, social facilities and affordable housing) is most of the time acquired through land readjustment (LR), and incidentally through expropriation.[1] Since the 1980s, most urban development in Spain, except for some large, strategic infrastructure (highways, railways, airports, harbours, large industrial areas, etc.) and urban plots already provided with infrastructure, has successfully been undertaken through LR. This includes not only urban expansion in greenfield sites, but also urban renewal and densification of existing urban areas: almost all old deteriorated industrial areas located within urban areas and many older residential areas have been redeveloped through LR, most of the time into residential uses.

This chapter gives an insight into the legal basis and execution of compulsory acquisition in urban densification in Spain.

The Spanish institutional context

To understand the working of compulsory acquisition in Spain, the institutional context needs to be explained first.

Spain is a decentralized state, with a system of constitutional parliamentary monarchy. From the 1980s onwards, 17 Autonomous Communities (*Comunidades Autónomas*, from now on 'regions') came into being, each of them with its own parliament and executive power. The decentralization of competences to the regions is such that Spain became de facto a federal state. Each region is divided into one or several provinces (*provincia*, of which there are in total 50), and each province into municipalities (*municipio*, of which there are in total 8,124). The governmental layers (central, regional, provincial and municipal) are organized hierarchically: the planning policies and plans of the lower governmental layers are subordinated to the law, policies and plans of the higher layers. Of these four governmental levels, provinces have the least competence and weight.

The distribution of legislative competences between the central government and each of the regions is based on the principle of symmetrical division of competences. Both the central and the regional governments produce 'formal legislation' (acts) and 'legislation in a material sense' (decrees and regulations). There are exhaustive lists of competences, some of them reserved exclusively for the central government, some of them exclusively for the regional government. But there are also policy fields in which the central government has the competence for making 'basic' legislation, and the regional government that for developing and detailing this basic legislation. In complex policy fields such as planning, central and regional competences are intimately related to each other. Within the planning field, the central government has the exclusive competence for property law, and the competence to issue basic legislation on common administrative proceedings, land and real estate appraisals and environmental protection. Regions can, within limits, detail and develop this basic legislation. As a result, the central government has, for example, important competencies for national infrastructure and the exclusive competence for expropriation law. The regions have exclusive competencies for practically the rest of planning law, including LR. Because regional legislation developed since the 1980s from a common original body of older central legislation, LR shares presently some basic similarities in all the regions (Muñoz Gielen, 2010, 105–106).

Neither legislation on expropriation (competence of the central government), nor on LR (competence of the regions) distinguish between urban expansion in greenfield sites and urban renewal or densification in existing urban areas. The procedures and conditions of expropriation and LR described in the next sections are thus common for both urban expansion and densification.

Urban renewal and densification through expropriation or through land readjustment

When public bodies prescribe land-use regulations that do not match the actual use and building possibilities (i.e. when public bodies prescribe redevelopment, e.g. when a new land-use plan foresees other uses and buildings than the existing ones), public bodies are entitled to use their statutory powers to acquire the property rights needed for implementing these new land-use regulations. In Spain, use and building possibilities are prescribed in a legally binding land-use plan. Municipalities have the competence for making this plan. There are two sorts of relevant plans at the municipal level: General Municipal Planning (*Planeamiento General Municipal*) and Detailed Planning (*Planeamiento de desarrollo*) (Parejo Alfonso and Blanc Clavero, 1999, 111–113, 264–286, 317). The General Land-use Plan (*Plan General de Ordenación Urbana, PGOU*) belongs to the first sort and covers the whole municipal territory. Each municipality must approve one of these plans, which are substituted by a new one every 5–20 years. Small municipalities have a simpler form (*Normas Subsidiarias*). The region approves them definitively. Detailed planning documents (Special Plans – *Plan Especial*, Partial Plans – *Plan Parcial* and Urban Renewal Plans – *Plan de Reforma Interior*) are meant for detailing afterwards the approved General Plans for specific sites, and they can modify their contents. Commercial developers can submit a Partial Plan and an Urban Renewal Plan as well as public agencies. It is, however, always the municipality which must approve such plans (Betancor and García-Bellido, 2001, 108).

Municipalities are thus entitled to prescribe in different sorts of legally binding documents land-use and building possibilities that do not match the actual use and building possibilities, and this gives them statutory powers to acquire the property rights necessary to implement the new use and building possibilities. The possibility of prescribing urban densification, and thus of deploying statutory powers to compulsorily acquire the necessary properties is hence not limited to specific sorts of sites. Compulsory acquisition can happen through two instruments: expropriation or land readjustment.

Expropriation

Public bodies, when confronted with the need to acquire land, are free to choose to apply their expropriation powers. This is the case both when public bodies want to construct public infrastructure alone and need the land under this infrastructure (most of the time large infrastructure like highways, railways, airports, harbours, etc.), or when they want to implement legally binding land-use plans, including the local public infrastructure linked to this plan (this is called *urbanización* in Spanish, 'urbanization': the provision of roads, green infrastructure, parks, parking, plots for social facilities, affordable/social housing, etc.), most of the time on-site (within the specific development site), but sometimes also off-site (outside the specific development site). In the case of construction of public infrastructure alone, a specific infrastructure decree or a legally binding land-use plan serves as legal basis for expropriation. In the case of implementation of legally binding land-use plans, the approval of this plan suffices as legal basis for expropriation.

When public bodies expropriate to construct public infrastructure alone, they can expropriate only the land needed. When public bodies expropriate to implement legally binding land-use plans, they can expropriate the entire development area – thus not only the land needed for public infrastructure, but also the land that will result in serviced building plots.

Expropriation is mainly used by central and regional public bodies for the sake of construction of specific large public infrastructure with a national or regional relevance, but not so much for obtaining infrastructure with a local relevance. There is also a longstanding tradition of central governments expropriating land in order to provide social/affordable housing (the 1911 Act on 'Cheap Houses' started this). The regions continued this practice from the 1990s onwards, but they have been contested by jurisprudence, on the basis of doubt about the public interest of the expropriation.

Local infrastructure is most of the time obtained by municipalities through LR instead of expropriation. There might be still many municipalities prescribing expropriation of land for the sake of local public infrastructure (most of the time located off-site of specific development sites), but they are often not successful in this and there seems to be a tendency towards using LR instead (Gozalvo Zamorano and Muñoz Gielen, 2017). Urban renewal and densification is most of the time the result of either owners of already serviced building plots redeveloping their plots (i.e. demolishing the existing buildings and building higher ones), or the result of municipalities using LR to reorder the property boundaries and provide new public infrastructure (or renovate the existing one). Urban renewal and densification are thus only marginally the result of expropriation.

Land readjustment

The first LR regulation was introduced in 1956 by the central government, and since the 1980s each region has developed the central regulation into its own regional LR regulation. Although the provision of public infrastructure necessary for the implementation of legally binding land-use plans, also called 'urbanization' (*urbanización*) in Spain, is still considered a public task, the planning system gives public bodies the statutory powers to allocate the costs of its implementation to the landowners through the principle of 'equitable redistribution of benefits, costs and duties' (*redistribución equitativa de beneficios y cargas*) of urban development. Land value increase and costs of development are proportionally distributed between all landowners according to the area of their original property. Land value increase thus serves to compensate landowners for taking responsibility for the unprofitable parts of urban development, i.e. for providing the public infrastructure and ceding the needed land (which includes compensating previous users, tenants and other title holders).

The planning system specifies in a LR regulation several instruments for operationalizing the redistribution of betterment and development costs. First, public bodies can allocate onto landowners the obligation not only of sharing betterment and costs, but also of organizing together the readjustment of the properties and the infrastructure provision (*Sistema de Compensación*). The LR regulation prescribes how landowners are obliged to jointly create a development organisation (*Junta de Compensación*). In order to follow this procedure, the owners of at least a specific percentage of the surface of the development area (in some regions 60 per cent, in others 50 per cent) must agree to do so and undertake the following steps. After the public body in charge (most of the time the municipalities) approves the legally binding land-use plan, the *Junta* must submit a Land Readjustment Project (*Proyecto de Reparcelación*, which includes a proposal for the readjustment of the property boundaries and for the allocation of development costs and resulting serviced building plots among the landowners) to the same public body. After approval of this project, the landowners who did not agree with the plan have two options. First, they can agree to participate in the readjustment, and thus voluntarily deliver their land and pay their share of the development costs to the *Junta* (*in pecunia* or *in natura*, i.e. in money or in serviced building plots). Second, there is in some regions the possibility for landowners of requesting expropriation of their land instead of participating in the readjustment (see section below). The public body in charge proceeds then to expropriate their land in favour of the *Junta*, which will have the legal status of beneficiary. In case landowners do not opt for any of the above alternatives, they are subject to compulsory land readjustment: their properties are reparcelled, they must pay their share of the costs (they can choose to pay *in pecunia* or *in natura*) and they receive a serviced building plot as compensation. If they don't pay, the municipality gives to the *Junta* some of their building plots as payment *in natura*. Subsequently, the *Junta* arranges the engineering works, provides the site with the needed public infrastructure, redistributes the resulting serviced building parcels among the landowners, both those that agreed with the Land Readjustment Project and those who didn't. The former landowners (except those who requested expropriation) thus obtain serviced building plots that are much smaller than their original plots but represent a much higher economic value.[2] Finally, the *Junta* transfers the public infrastructure to the public body in charge, and then the *Junta* dissolves itself.

The planning system includes in addition two major instruments that give public bodies statutory powers to intervene directly. First, the LR regulation gives public bodies statutory powers to overrule the landowners and take the lead by providing the infrastructure and applying compulsory land readjustment (*Sistema de Cooperación*). This is the case, for example, if no majority of landowners join the *Junta*, or in case they do but afterwards they do not proceed accordingly and delay the readjustment. The LR regulation also gives to public bodies the powers to apply compulsory land readjustment right away, from the beginning, without the need to try first the voluntary way of the *Sistema de Compensación* described above. Public bodies can thus apply compulsory land readjustment both in case landowners do not agree to follow voluntary land readjustment and also in case public bodies decide so from the beginning. In all cases, compulsory land readjustment means that properties are reparcelled, landowners must pay their share of the costs (they can choose to pay *in pecunia* or *in natura*) and they receive a serviced building plot as compensation. There is thus no formal need of agreement with any percentage of landowners.

The second major instrument for public bodies to intervene directly is Expropriation law, which as mentioned in the previous section, gives public bodies statutory powers to expropriate the entire development site (*Sistema de Expropiación*).

Some regions introduced in the 1990s a third major instrument for public bodies to intervene directly: to select a third party (the urbanizing agent, *agente urbanizador*) as implementing agency, i.e. as being responsible for carrying out the infrastructure provision and readjustment of the properties. This third party was often seen as necessary because landowners often did not cooperate, and public bodies were reluctant to deploy a more active role. This need was felt especially in those areas with traditional fragmentation of landownership. The region of Valencia was, in 1994, the first one to introduce this third option. The 2007 Planning Law of the central government introduced this option too, but its operationalization depends on regional legislation. Some regions have thus effectively introduced this third option, and others not. Regions with larger landownership patterns, like Madrid, or regions with more conservative approaches to landownership, like Galicia, however still support mainly voluntary LR. In those regions that implemented the Valencia model of LR, the implementing agency does not necessarily have to own property in the development area, and municipalities can select it in a public tender, in which any other party (including the landowners) can participate. In the first versions of this legislation, which prevailed until beginning of the 2010s, public bodies did have the power to freely choose that urbanizing agent which proposed the most suitable plan, regardless of whether landowners supported the agent or not. Only where landowners submitted an alternative plan that performed as well as the agent's plan, did public bodies have to select the owners' plan. In practice, there was thus no formal obligation to negotiate with the landowners and this gave public bodies large room to decide which plan to approve. As a result, the position of landowners was considerably weakened. However, this being said, it is also true that in many cases the urbanizing agent (who often was a developer owning part of the land or at least acting in agreement with part of the owners) did negotiate support for their plan. The reason for that is that public bodies, specially municipalities, although formally not obliged, very often required urbanizing agents to obtain a minimum support of landowners in order to avoid social conflict and opposition. In the 2010s, the position of landowners has gradually been formally reinforced by acknowledging them preferential treatment: since then, the urbanizing agent with most support among landowners is given preference. However, the agent's proposal must still fulfil minimum requirements, and in case it doesn't, or in case landowners do not support any plan, public bodies still have the statutory powers to select that urbanizing agent that submits a plan and fulfils all requirements.

The procedure for the compulsory LR looks in all regions more or less as follows (Parejo Alfonso and Blanc Clavero, 1999, 454–461):

- First public consultation about the legally binding land-use plan and, in those regions in which an urbanizing agent can be selected as the implementing agency, a draft of the public infrastructure, of the cost estimation and of the development agreement (the so-called *Programa*).
- Approval plan: the public body in charge (in case of a municipality, the local council, *Pleno del Ayuntamiento*) approves the legally binding land-use plan and, in those regions with the possibility of nominating an urbanizing agent, also the *Programa*.
- Submission Land Readjustment Project: after the approval of the legally binding land-use plan, the implementing agency in charge of LR (the *Junta* of landowners in the *Sistema de Compensación*, the public body in the *Sistema de Cooperación* or the urbanizing agent in some regions), must submit a Land Readjustment Project.
- Second public consultation: the Land Readjustment Project is announced generally in newspapers, and announced individually only to the registered

titleholders, that is those owners that are registered in the Property Register (*Registro de la Propiedad*).

- Application to the Property Register of a note that certifies who are the titleholders and which charges exist on the property (*certificado de dominio y cargas*). At the same time, the Register notes the initiation of the readjustment procedure.
- Additional hearing in case there are titleholders that have not been formally addressed because they were not inscribed in the Property Register. Also, an additional hearing for the affected persons in case of modification of the Land Readjustment Project.
- During the public consultation, affected persons can claim compensation from each other. The owner of the freehold of the land has to compensate the titleholders of other rights on their property that have to disappear. In the case of tenants having to leave their house, tenants legislation applies. This may delay the procedure.
- During the public consultation, owners can negotiate, i.e. offer, or demand from each other, shortages and surpluses of building rights (*defectos y excesos de aprovechamiento adjudicado*). It is not always possible to allocate all the building rights of each landowner on to a particular building plot. When there are shortages or surpluses, and it is not possible to divide the new plot between several owners (*pro indiviso*), owners have to buy or to sell the remaining rights. In the case that negotiations among landowners are not successful, the Land Readjustment Project proposes a unitary price for the building rights at which the implementing agency in charge of LR (the *Junta* of landowners in the *Sistema de Compensación*, the public body in the *Sistema de Cooperación* or the urbanizing agent in some regions) offers to buy the surpluses of building rights. If the owners and implementing agency do not agree with this price, the landowners can propose another price. If there is still no agreement, only the judicial avenue remains for landowners.
- Approval of Land Readjustment Project: the public body in charge approves the Project.
- Publication, individual announcement to the affected persons and registration of the approved Land Readjustment Project in the Property Register. This registration implies the legal transfer of property, both of the serviced building plots (which go to the landowners) and of the public infrastructure (which goes to the public body responsible for the specific infrastructure, most of the time the municipality).

The procedure for voluntary LR is simpler: the Land Readjustment Project has to be endorsed by all the affected persons (landowners and other titleholders). Once the project has been submitted, the public body in charge decides right away: there is no period of public consultation.

Almost all urban development in Spain (especially greenfield, urban extension; but also urban renewal and densification) has been developed through LR

(Muñoz Gielen and Korthals Altes, 2007, 69; Muñoz Gielen, 2010, 125–128). The rest is implemented through expropriation or voluntary land acquisition: strategic developments (e.g. large industrial sites, some public facility complexes, social housing and other individual cases), large infrastructure (e.g. highways, railways, airports, seaports, etc.), and some urban renewal projects with strategic importance, mostly in the oldest or more distinctive urban historic areas. Urban densification in plots already provided with public infrastructure that do not require property readjustment and/or the construction of new public infrastructure is most of the time implemented by their owners, without need of expropriation or LR. Thus here voluntary land acquisition prevails as way of urban densification. LR is, most of the time, applied by municipalities. Often, landowners voluntarily join the readjustment procedure, but the possibility of applying a compulsory track is considered as being essential for the effectiveness of LR (LR often involves hundreds of small and large landowners).

Expropriation versus land readjustment

As mentioned before, there exists the possibility that landowners not willing to participate in the land readjustment are expropriated instead. The municipality proceeds then to expropriate their land in favour of the implementing agency in charge of LR (the *Junta* of landowners in the *Sistema de Compensación*, the public body in the *Sistema de Cooperación* or the urbanizing agent in some regions), which will have the legal status of beneficiary. That is, the municipality proceeds to carry out the expropriation procedure but the implementing agency pays the compensation and receives the property of the expropriated plots.

This possibility of expropriation is included for example in the LR regulation of the region of Andalucía. However, other regions try to avoid the possibility of expropriation and promote in their LR regulations the obligation to participate in LR (that is, to be compensated with final serviced building plots, not with a priori monetary payment). This is the case in the region of Basque Country, which foresees the possibility of applying compulsory LR without the possibility for landowners to request the expropriation of their land.

Compensation in expropriation

Before 2007, compensation for expropriated property was based on its future value, thus including any subsequent development value. However, from the end of the 1990s until 2007 compensation sums were often lower than were the profits of participating in LR, mainly because at that time market prices rose so fast that appraisals of compensation sums were soon outdated.

The 2007 Planning Law of the central government (*Ley del Suelo*) introduced however a new appraisal method. Since this law, compensation has increased along with the fulfilment of the obligations in the land readjustment procedure. When land is expropriated before any step has yet been undertaken (the Land

Readjustment Project has not yet been approved and thus neither the infrastructure has been provided, nor the property boundaries readjusted), compensation is based on the former use and building possibilities of the land. Most of the time this should lead to relatively low compensation sums because they do not include the future use and building possibilities. As obligations are fulfilled, compensation gradually includes the future use and building possibilities, thus leading to higher compensation. This appraisal system is intended to stimulate participation of landowners in land readjustment, and is related to the principle, included in the Spanish Constitution, that building rights do not automatically belong to the landowners. Landowners must first fulfil their obligations and can then, and only then, profit from land value increase.

It seems however that in practice this novelty is not having the intended consequences. First, in 2007 the economic crisis started in Spain, which till recently has diminished land values. There is thus no or much lower land value increase, and compensation sums are thus not far away from the land value based on future use and building possibilities. Second, the Constitutional Court declared unconstitutional part of the appraisal rules, which in practice hampers the effective application of the appraisal rules of the 2007 Act. And third, the Supreme Court has been ruling for the last five years in favour of landowners, recognizing the highest compensation values (Lora-Tamayo Vallvé and Parada, 2016).

Here follow two cases that illustrate how LR is used in urban densification.

Illustrative case, micro: Músico Chapí, Valencia

In December 2003, the Local Council of the Municipality of Valencia selected developer *Asociación de Interés Urbanístico (A.I.U.) PAI Músico Chapí* (an association of the landowners of 46 per cent of the development area and a property developer who bought a large share of the land) as urbanizing agent in *Músico Chapí*. The urbanizing agent is since then thus in charge of the LR-procedure, i.e. of readjusting the properties and providing the public infrastructure. *Músico Chapí* is a 2.9 ha site located to the south of the City of Valencia which was initially divided into 41 parcels. The 1988 General Land Use Plan already had previously rezoned the former uses (peripheral deteriorated urban area, with housing, small industries and some undeveloped land) into residential (32,806 m^2 floor space), public infrastructure and green zones. Afterwards (but before the public tender in which the urbanizing agent was selected) the Municipality modified the 1988 General Plan to substantially diminish the area of buildings (from 32,806 to 23,801 m^2 floor space) and to increase the public infrastructure (from 58 per cent to 62 per cent of the total surface) (see Figure 7.1).

In addition to the legal minimum of obligatory, non-negotiable statutory contributions (ceding 62 per cent of the development area and building public infrastructure – roads and green zones), the urbanizing agent negotiated a complementary contribution to deliver social/affordable housing (15 per cent of the total new building volume). The property developer who bought a large share of

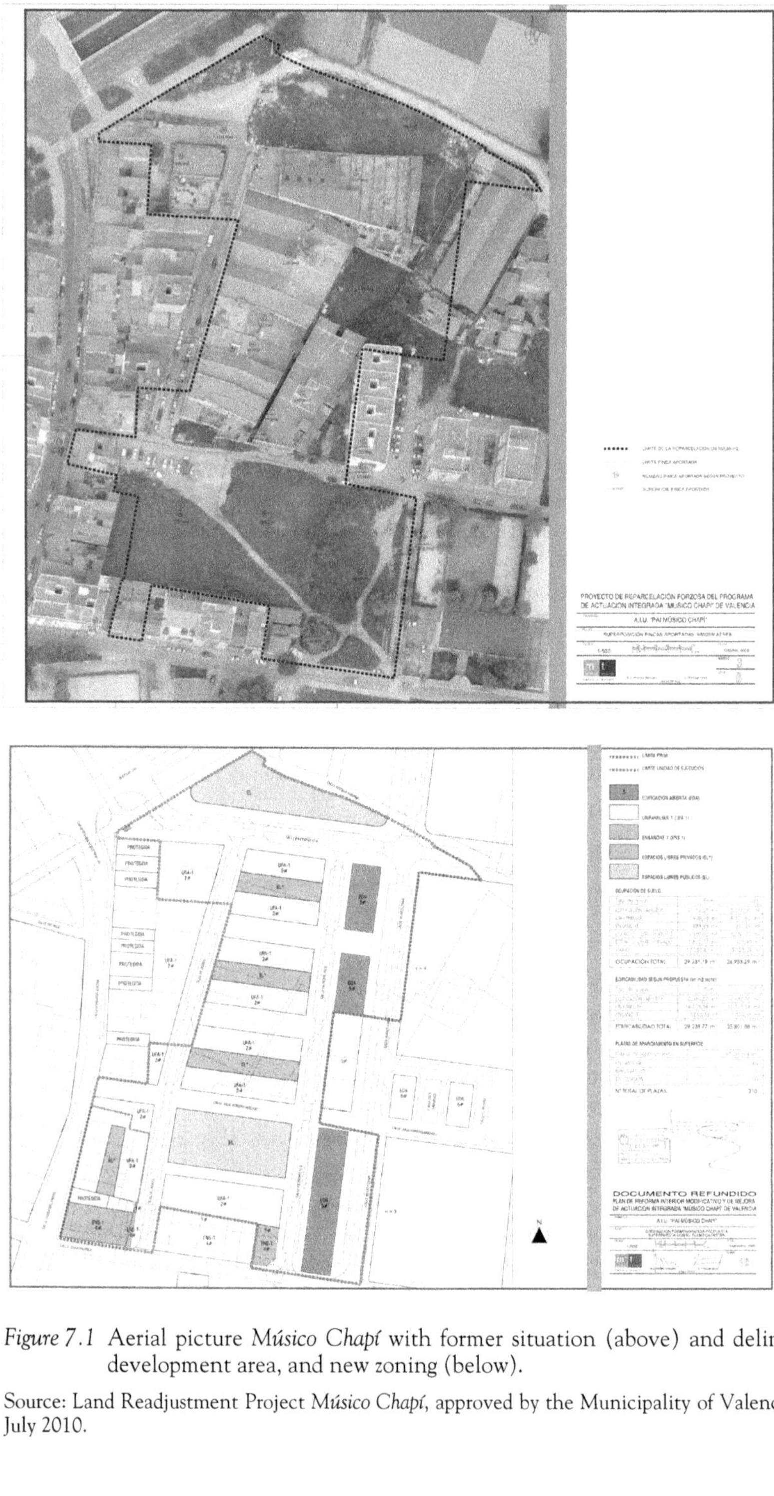

Figure 7.1 Aerial picture *Músico Chapí* with former situation (above) and delimitation development area, and new zoning (below).

Source: Land Readjustment Project *Músico Chapí*, approved by the Municipality of Valencia on 30 July 2010.

the land did not charge this contribution to the rest of landowners. The Land Readjustment Project was approved in July 2010. Some existing buildings have been demolished but the works have not proceeded further because an old factory chimney with historic value has been discovered. The municipality is elucidating now whether the chimney has to be protected, and if so, how the zoning must be modified and who will bear the extra costs (these could lead to more negotiations with the urbanizing agent about additional costs) (Muñoz Gielen and García Pastor, forthcoming).

Illustrative case, macro: Barcelona @22

Since 2000 the Municipality of Barcelona has modified General Municipal Planning to allow the renewal of 130 ha of the district *Poble Nou* (an old nineteenth-century industrial area) into a mix of new technological industries, offices and housing. The Municipality prescribed the reconstruction and substantial improvement of the road infrastructure and public space, including modern underground communication networks (internet, telephone, etc.), garbage removal facilities and central air conditioning (total cost €180 million). It also prescribed the free cession of buildings for public facilities and of land for new public space, green facilities and social housing. The municipality also prescribed the refurbishing and conservation of 104 historic buildings and constructions, mostly industrial.

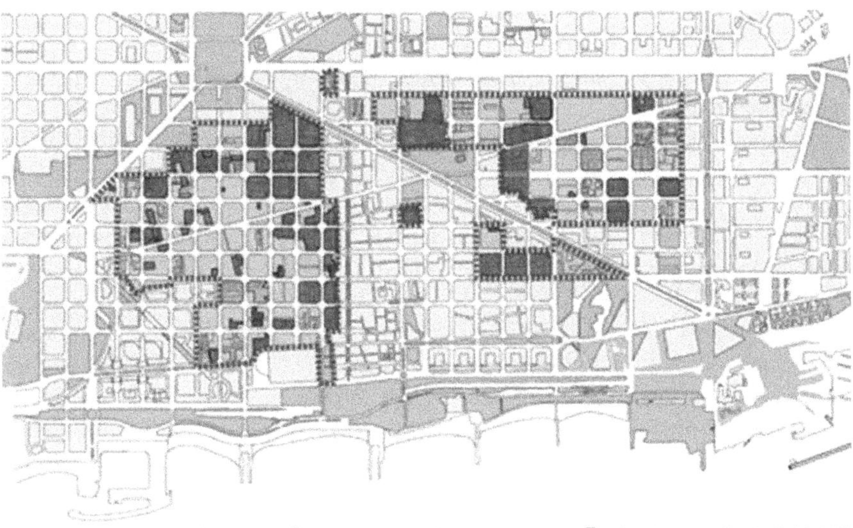

Figure 7.2 Urban renewal area @22. In dark grey, definitively approved LR projects; in medium grey, provisionally approved projects (information up to December 2011).

Source: www.22barcelona.com.

In December 2011, 47 per cent of the area (625,000 m^2 of land) was developed or in the process of development. About 35 per cent of the area was developed or is in the process of development through land readjustment. Landowners, most of the time through land readjustment and sometimes directly, developed serviced plots with a building capacity of 1,770,000 m^2 floor space, and ceded for free to the municipality 200,000 m^2 of land for public spaces and 150,000 m^2 of floor space for public facilities and social housing (1,300 social housing units were built, and 500 were under construction or already had a building permit). In addition, landowners contributed €45 million for the refurbishing and improvement of the road infrastructure and public space (www.22barcelona.com).

Conclusions

In general, the approval of a legally binding land-use plan suffices as the legal basis for two statutory instruments that allow public bodies to compulsorily acquire that property which is necessary for implementing whatever is foreseen in the legally binding land-use plans, and thus also for urban renewal and densification. These instruments are the expropriation of land, which is only applied in specific cases, and land readjustment (LR). LR is the most common way for organizing urban development in Spain, especially in urban expansion, and to a lesser degree also in urban renewal and densification. In Spain, land can thus be compulsorily acquired not only through expropriation, as is the case in many other countries, but also through LR. In general, expropriation is the least popular, among other reasons because of conservative (pro-owner) standpoints in jurisprudence. LR offers a more stable, financially feasible and traditional operational framework.

The development industry plays a central role in facilitating, encouraging and promoting the compulsory acquisition (sometimes through expropriation, most of the time through LR) of that property necessary for urban development, i.e. urban expansion and also urban renewal and densification. The development industry, if it owns part of the land, is involved as a precursor of LR: it often initiates the LR procedure and obtains the support of public bodies to force LR on those landowners opposing it. When public bodies expropriate landowners not willing to participate in LR, the development industry becomes the legal beneficiary of expropriated land. It must be noted too that, compared with other countries, landowners are more intensively involved in the implementation of legally binding land-use plans thanks to the possibility of joining the development through LR. Often landowners are the precursors themselves of LR, i.e. they become an element of the development industry. There is a consolidated practice of mortgage loans, with the final expected building serviced plots as collateral, that allow landowners to easily obtain the necessary initial capital against low interest rates. In some regions, the involvement of the development industry not linked to landownership (thus not owning much land) has been more intensive because there is the possibility of implementing LR without owning the land. In these regions, most notably in the region of Valencia during the boom times of the housing market (1997–2007), a broader spectrum of the development industry was able to invest in urban development, leading to a remarkable increase in investment in urban expansion and also in urban renewal and densification.

Landowners whose property is acquired in order to make urban development possible are compensated with an upfront monetary payment in case of expropriation, or with final serviced building plots in the case of LR. Compensation sums in the case of expropriations were based until 2007 on the future, more profitable land use and building possibilities, but in practice the fast escalating market prices from 1997 to 2007 led to landowners preferring LR instead of expropriation: market values of final serviced building plots often exceeded expropriation sums up front. In 2007, the central government modified the appraisal rules, trying to reduce compensation sums and stimulating the participation of landowners in LR, but in practice not much seems to have changed.

Notes

1 Expropriation is equivalent to 'eminent domain' in the US and to 'compulsory purchase' in the UK.
2 When urban renewal and densification takes place in urban areas with a high density (apartment blocks and high rise offices, floor area ratio around 1), landowners must cede for free an average of 60–80 per cent of the total surface for public infrastructure, and in peri-urban sites with low densities (single-family dwellings, under 1) 30–60 per cent (Gozalvo Zamorano and Muñoz Gielen, 2017).

References

Betancor R A and García-Bellido J (2001) Síntesis general de los estudios comparados de las legislaciones urbanísticas en algunos países occidentales, *Ciudad y Territorio-Estudios Territoriales*, 33(127), 87–144.

Gozalvo Zamorano M J and Muñoz Gielen D (2017) Non-negotiable developer obligations in the Spanish land readjustment: an effective passive governance approach that 'de facto' taxes development value?, *Planning Practice & Research*, 1360–0583 (online). http://dx.doi.org/10.1080/02697459.2017.1374669.

Lora-Tamayo M and Parada J R (2016) *Derecho Administrativo III. Bienes Públicos y Derecho Urbanístico*. Madrid: Open Ediciones.

Muñoz Gielen D (2010) *Capturing Value Increase in Urban Redevelopment*. The Netherlands: Sidestone Press.

Muñoz Gielen D and García Pastor M (forthcoming) Transparency and evolution in the use of negotiable developer obligations in urban development in Spain.

Muñoz Gielen D and Korthals Altes W K (2007) Lessons from Valencia: separating infrastructure provision from landownership, *Town Planning Review*, 78(1), 61–79.

Parejo Alfonso L and Blanc Clavero F (1999) *Derecho Urbanístico Valenciano. (Análisis de la Ley Reguladora de la Actividad Urbanística)*, 2nd edition. Valencia: Tirant lo blanch.

8 Densification, dispossession and disposable lives

A case study of urban space production through the Expo in Shanghai

Yunpeng Zhang

Introduction

Over half of world's population nowadays live in cities and it is expected that this figure will increase to 60 per cent by 2030 (United Nations, 2016). Planetary urbanization has brought a host of challenges: housing unaffordability, environmental degradation and social inequalities, to just name a few. In order to address them, urban densification has been advocated by political elites and planning professionals as a potentially effective tool. However, it has varied success in practice due to different political-economic configurations and community resistance (Searle and Filion, 2011; Poppe and Young, 2015).

This chapter examines densification and its dispossessory arrangements in Shanghai. It does so through an in-depth case study of the socio-spatial transformation caused by the World Expo 2010. Densification is not only one justification for the site choice of the Expo but also a crucial means to balance the finance for land clearance and property dispossession through compulsory purchase. According to official statistics, the Shanghai government evicted more than 18,000 registered households and 272 businesses from the Expo Park within Shanghai's urban core and over 5,000 households from the site of resettlement housing projects in Shanghai's urban fringes, excluding renters and those displaced who lost their homes to intensive infrastructure development and city beautification projects. Displacement and resettlement alone cost municipal finance 30 billion RMB. The funding for it was largely reliant on land redevelopment of the same site after the six-month global spectacle.

In spite of this sheer scale of dispossession and displacement, the project was pushed through with limited resistance. This is not because there was a consensus over the intended outcomes of densification, although the Shanghai government and its cultural allies dedicated a considerable amount of energy to doing so. After winning the bid for hosting the Expo in 2002, the Shanghai government organized a city-wide discussion in 2003, bringing together ordinary citizens and experts from different fields and sectors to construct a shared vision on the opportunities and challenges brought by the Expo (Contemporary Shanghai Institute, 2003). However, given the *post hoc* nature of these debates, by design they were not intended to question the necessity of hosting the Expo or redevelopment agendas. They were more to manufacture legitimacy for anticipated huge public spending, compulsory acquisition and forced evictions.

The absence of resistance, this chapter argues, results from the dispossessory arrangements and distributive politics. These arrangements constructed a hierarchized valuation of human bodies, which undermined the potential for organized collective resistance and deprived the displaced residents of the opportunity to steer the process of socio-spatial changes and co-produce denser but more inclusive urban spaces. For this reason, this chapter maintains that densification in the case of the Expo is incapable of addressing development challenges and, on the contrary, has contributed to solidifying existing socio-spatial inequalities. It justifies and relies on compulsory purchase, an act imbued with state power and state violence, but the intended goals of this planning tool are compromised by the vested interest of agents of capital and state in revenues derived from land and real estate development on the one hand and by the exploitative, authoritarian and sometimes violent practices of compulsory acquisitions on the other. Without changing the dispossessory logics and arrangements in land (re) development, it is unlikely that densification will ever achieve its intended outcome.

This chapter will utilize a collection of policy papers, newspaper articles and interviews. They were collected during my fieldwork for a larger research project on the processes and the lived experiences of residential displacement caused by the World Expo in 2012. It will first contextualize the Expo-induced socio-spatial transformation against the backdrop of dispossessive urbanism in China. It underscores the crucial role of dispossession in China's political economy and discusses prevailing ways in which dispossession is organized. It does so because any planning interventions that involve land use changes simply cannot be discussed in isolation from the land-based accumulation strategy. This is followed by a dissection of dispossessory arrangements in the Expo-induced land clearance. It will highlight the political calculation that differentiates and orders human bodies, and discusses its divisive effects. In conclusion this chapter calls for a more participatory approach to land use planning and greater care in dispossessory arrangements.

Dispossessive urbanism in China

Dispossession is identified to be a crucial mechanism for the emergence and reproduction of capitalist relations (Rossi, 2013). It also lies at the heart in the making of the Chinese variant of capitalism. Since the mid-1990s, China gradually shifted from rural industrialization towards a land-based urban-centric accumulation (Hsing, 2010). Urbanization as city building has been pursued by the Chinese state to accumulate capital and power. This shift also entails important changes in the meanings of land. Increasingly, its function as a commodity and a financial asset has been prioritized. Revenues from land and real estate development account for an important part of municipal finance after the fiscal and tax reform in 1994 (Ding and Lichtenberg, 2011). Continuous land dispossession and commodification is essential to sustain the economic growth and finance public spending.

Of course, in making this claim I am not suggesting that capital accumulation is the only logic that induces dispossession in China. Based on his research in rural China, Webber (2008) identified several extra-economic triggers of dispossession, including environment modernization, ethnic politics, nation building and personal motives. But, given the reliance of public finance on land, economic motive has been a dominant one. It is also very likely that these instances of dispossession for extra-economic purposes are funded by revenues generated from land development.

Dispossession is a process dominated by the state, and in most cases, through compulsory acquisitions. This owes much to the state dominance in land administration in general. Within China's dualistic land tenure arrangement, all land in cities is owned by the state. While land in rural areas is under collective ownership in law, the question as to who represents the collective was not clear in the legislature, leaving ample room for grassroots agents to abuse power and pursue personal interest (Ho, 2001). Moreover, due to administrative subordination of rural regions to urban authorities, rural land use is significantly influenced by the latter. Such influence is further consolidated by the legal arrangement, establishing the state as the monopoly agency in converting collectively owned rural land for urban centric development. Land categorized and explicitly earmarked as basic farmland in land use plans can only be converted for projects of national or regional significance with the approval of the state council or provincial-level governments. But in practice, sub-provincial authorities often bypass it through the Link Policy (see below and Chapter 9), exchanging rural land of equivalent sizes in different locations.

State dominance is also exercised through planning control, which often lays the conditions for compulsory purchases. In the Chinese context, planning is to serve the interests of pursuing rapid economic growth and legitimating the expansion of territorial control by the state (Abramson, 2006; Wu, 2015). Planning restrictions can contribute to neighbourhood decline (Wu, 2016). This is particularly so in neighbourhoods with plenty of self-built housing. Residents are discouraged from improving their housing conditions by altering built structures, even if these modifications are intended to accommodate changing housing needs. Considering the shortage of affordable housing, these changes may to some extent be tolerated by the state. Yet, they can also lend legitimacy to discourses stigmatizing such improvisations as unruly practices, an alibi for property dispossession and displacement (Pan and Liu, 2011). In these events, such building modifications are usually classified as illegal, excluded or unfairly compensated.

Another important means through which the state dominates dispossession is the law. Lubman's famous avian metaphor of *Bird in a Cage* aptly captures the contradictions inherent in the post-Mao legal modernization project (Lubman, 1999). Law is conceived as an instrument by the party elites – the bird keepers – to promote economic growth and manufacture legitimacy for their durable dominance without subjecting these elites to the legal constraints. This contradiction is clearly evident in the sphere of property dispossession and residential displacement. Regulations have been revised in a way that is consistent with the imperatives of maximizing economic returns through land development and expanding the role of the market while maintaining political control.

Owing to the rural-urban divide and the resultant dualistic land ownership, regulations with regard to property expropriation also differ in terms of procedures, compensation standards and legal remedies. By way of comparison, regulations protect the interests of urban property owners more than rural property owners (People's Republic of China, 2004; 2007; State Council, 2011). This is unsurprising given that acquiring land cheaply from rural property owners and increasing development density is critical to ensuring the profitability of land development. But it is quite disturbing because most instances of dispossession take place in the countryside where the law offers very weak protection of existing land users and tends to compromise for the imperative of economic growth.

This said, it is noteworthy that the definition of public interests, a precondition for compulsory purchase, is kept deliberately vague and somewhat contradictory in Chinese legislation in order not to completely deprive local governments the incentives and channels to pursue economic growth. Article Eight of the current regulation on Acquisition and Compensation for Housing on State-owned Land (effective in 2011) lists eight situations that meet the public interest test. Within them, economic development is considered to a sufficient justification, assuming what is good economically is good for the public. Similarly, it also qualifies redevelopment of areas earmarked by urban planning due to poor housing quality or insufficient infrastructure to be of public interest, despite that such redevelopment often smuggles in for-profit real estate development and original residents gain marginally from the process as they have to find shelter elsewhere.

Therefore, urban property owners are equally susceptible to legal violence. Under current regulations, property owners and existing land users rarely participate in the planning of development projects. As a consequence, they can neither challenge the development plan nor claim their right to stay put once a decision has been made. In fact, with the establishment of land banks, satellite state institutions that expropriate land and properties with or in many cases without pre-defined projects so as to consolidate the state's monopoly over the primary land market and maximize land's exchange value, it is plainly impossible for Chinese citizens to exert meaningful control in the process. With little prospect of staying put, compensation and resettlement become central to the claims of property owners.

While China's land-based accumulation has largely avoided causing large-scale physical dislocation, resettlement remains secondary to displacement and land clearance in this process. Arrangements of resettlement have shifted from in-kind to monetary compensation and from on-site resettlement to off-site relocation. In the case of rural land and property expropriation, compensation usually comprises the residual values of properties, contribution to social insurances of rural residents, the value of agricultural outputs, and relocation costs as well as bonuses for moving out in time. Compensation for urban property owners, on the other hand, is higher but excludes social security contributions and agricultural output compensation. This is not surprising because urban residents usually engage in non-agricultural employment and they are already covered by social security tied with their urban resident status and employment. Thus, extending the coverage of society security to rural residents simply addresses institutionalized inequality rooted in the urban-rural divide rather than redistributes land revenues. It is also important to recognize that in both cases land value is excluded from calculation given that land ownership is separated from use rights and property owners have only use rights to land where their properties stand. Compensation is constrained not only by the law, deciding what can be compensated, but also by the interests of the state and its alliances for land development. It is possible for the to-be-displaced residents to hold out and bargain for better resettlement deals (Ho, 2013). And indeed, it is not uncommon that residents manage to squeeze more out of the agents of the state and capital through individual or collective actions (Shao, 2013). Nevertheless, this is contingent on power struggles in negotiations. As dispossession and displacement is often outsourced to companies, albeit with some links with state agencies, there is a tendency for these companies to minimize the payments to those displaced. Holding out then carries the risks of being stigmatized as selfish and greedy or being subjected to the extreme violence of forced evictions (Shao, 2013).

In appearance, the law offers a tool for citizens to voice their grievances and keep state power in check. But the lack of judicial independence and the politicized legal system allows the state to bend the rules to serve its own interests in practice (Lubman, 1999). Moreover, as Shih (2010) pointed out, the Administrative Litigation Law, which allows citizens to challenge state administrative decisions, restrains the court from challenging the validity or appropriateness of administrative decisions and limits the authority of the court to review legal conformity and procedural justice.

In claiming state dominance in land dispossession and development, I am not suggesting that the Chinese state is a coherent or monolithic actor. It is decentralized, fragmented and contradictory, along with the ongoing construction of the socialist market economy. This is evidenced by vertical conflicts along administrative hierarchies and horizontal tensions between different political factions, between different local states as well as between the imperatives of accumulation and legitimation (Lieberthal, 1995; Hsing, 2010; Fewsmith, 2013). These conflicts lead to competing visions and values over land and its use. Rapid dispossession over past decades has raised much concern over grain security and social stability. As a result, the central government has mobilized its administrative and political measures to restrain the scale of dispossession and discipline local state agents.

However, these measures do not alter fiscal arrangements between the central and local governments, a key cause for the latter to pursue a dispossessive land-based accumulation strategy. With little progress in reforming municipal finance, local governments under the pressure of inter-urban competition and career performance have looked for more ingenious ways to find land for real estate development within the political parameters set by the central government. There are three popular strategies. First, in order to co-opt those displaced and contain popular resistance, local governments frequently make economic concessions (Liu, 2016). Such concessions can be easily recovered through rising land values. Second, local governments have widely pursued farmland consolidation and restoration projects in rural areas (Wang *et al.*, 2010; Ong, 2014). Under the Link Policy (Chapter 9), the idea is to release land in better locations of the same sizes as reclaimed farmland for commercial development. Compulsory purchase of land and properties in rural areas is an important part of these projects. This is because rural settlements are usually constructed at a low density and plot ratio. By pushing up density of resettlement housing, it can potentially increase the supply of land for construction (Tan, 2014). In this way, local governments not only respect the authority of the central government but also pursue their own economic and political interests. A third strategy is to take advantage of national or regional development projects (Qu, 2012). Mega-events and transportation hubs fall into this category. Projects of these kinds often come with irrefutable legitimacy and exceptional rules in terms of land supply, land expropriation measures, compensation standards and financing strategies. Local governments often bundle them together with commercial development projects to profit from boosted land and property values in surrounding areas.

Rationalizing the hosting of the World Expo 2010 in Shanghai

Between 2008 and 2010, China hosted two global mega-events and one regional sporting event in three regional centres – Beijing (the Olympic Games), Guangzhou (the Asian Games) and Shanghai (the World Expo). As Horne (2007) remarked, for cities from developing countries, hosting mega-events is a rather questionable strategy given much more acute problems of poverty and inequality confronted by much of the population.

Like many host cities, the political and cultural elites mobilized a host of discourses to justify this decision to host the Expo. I categorize them into higher order ones such as nationalism, patriotism and modernization, and lower order ones such as urbanization, distributive justice and environmental sustainability. It must be noted that these discourses are not mutually exclusive. Neither does the higher/low ordering correspond to any spatial scales. The difference lies in the values they appeal to and the normative power they have. Another difference is that the lower order ones tend to concern more the decision of the Expo on the chosen sites while the higher ones concern more the symbolic benefits of the Expo. I maintain that it is the higher order ones that are more decisive in legitimatizing compulsory purchases and paralysing the agency of those displaced.

Before turning to examine the discursive construction of the Expo, I shall first briefly contextualize the Expo bid. As early as in the 1980s, elite politicians had already got in their mind to host an Expo in Shanghai. This was nurtured by a revised perspective of history, progress and modernity, eliminating class struggles from the ideology and concentrating on hard measurable facts of economic growth (Liu, 2009). However, limited by the infrastructure capacity of Shanghai, compounded by the economic crisis and the student movement at the end of the 1980s, Shanghai gave up the plan (Chen, 2011). When the idea was brought on the table again in the 1990s and translated into an actual bid at the turn of the century, facilitating economic growth through land dispossession and real estate development had been established as an orthodoxy. In 1992, Shanghai launched the notorious campaign, known as the 365 Plan, which classified inner city housing of 365 hectares as "dangerous" needing demolition. Although some of this housing required substantial improvement, many neighbourhoods with housing stock in good condition and low density were earmarked for demolition only to ensure profitability of gentrification projects (Ren, 2014). By the end of 2000, the Shanghai government celebrated the success of this 365 Plan. In total, more than 660,000 households, that is, about 2.5 million residents, lost their homes and relocated to suburban resettlement sites. This has resulted in a dualistic real estate market – a lower-end housing market for resettling those displaced in urban fringes or remote rural areas and a high-end residential and commercial market for the emerging middle class and the elites on the demolished and subsequently gentrified sites. The built area of Shanghai thus expands fast at low cost.

The initial choice of the Expo site was reflective of this logic of urban expansion. Three proposed sites were located in Shanghai's suburbia where land could be acquired cheaply. The inspiration for the final decision came from a planning exercise for the Expo park by a group of residents from a Paris-based planning studio during their summer camp in Shanghai (Chen, 2011). One team of students dismissed the assigned location by the Chinese planners and proposed to build the Expo park on Shanghai's waterfront in inner Shanghai, dominated by industrial landscapes. The boldness was appreciated by the Chinese planners. But the Chinese planners were aware of the difficulty in having the Expo site there because of the fragmented property rights and the costs of displacement and resettlement. It was the popularity of this proposal within Bureau International des Expositions, the governing body of the Expo, entertained by the Chinese politicians' commitment to win the bid at any cost, that forced the Chinese planners to evaluate the possibility of hosting the Expo on the waterfront. In less than four months, rationales for this new site choice were neatly reasoned and presented in the final bidding report. Similar to what Flyvberg (1998) noted in the Aalborg project, what we see here is that rationalization is produced whenever powerful elites need one.

Several justifications were offered to justify this new site choice (Editorial Committee of Shanghai Expo Planning, 2010). As Shanghai was envisioned to become an international centre of economy, finance, trade and shipping, industrial complexes and old unplanned housing that dominated the waterfront in the city centre were deemed obsolete for this changed strategic function of Shanghai. By moving out these industrial operations, the authorities wished to transform the waterfront for upper-end commercial housing, entertainment facilities and office towers. The Expo squared well with this overall strategy. Compared to building an Expo Park in the suburbs, which would cause further urban sprawl, hosting the Expo in the city centre would not only improve land use efficiency but also contribute to the functional upgrading of the city centre given the cultural facilities and expo venues that would be left behind after the event. Along with the spatial restructuring, the post-expo space was envisioned to attract a social group of a higher calibre. The original residents were expected to relocate to two giant resettlement sites constructed near or outside the outer ring of the city. Density of permanent population in the city centre would decline in general but increasing density in some areas was planned to ensure the return on investments. In the meantime, density in resettlement sites would be increased significantly. Relocation of businesses to industrial parks was also expected to bring more population flows and improve land use efficiency in surrounding regions. As a result of these spatial reconfigurations, it was expected that the Expo would improve environment quality and boost land values. Many industrial operations were the main source of pollutants. One planned legacy of the Expo was to increase green space and remedy the soil contaminated by industrial pollution.

To sell the event to the Chinese audiences, legitimatizing discourses were penetrated by nationalist sentiments. The Expo bid was put forward at a time when nationalist sentiment was paramount in China, resulting from not only the booming economy in the 1990s but also a series of geopolitical events such as the resumption of sovereignty of Hong Kong and Macau, the bombing of the Chinese Embassy in Belgrade by US troops and China's accession to the WTO. The Expo thus offered an important site to articulate state-sponsored nationalist sentiment. The temporal proximity between the beginning of China's humiliating century (1841) and the invention of the Expo (1851), and Shanghai's colonial background, provided the cultural elites with ample room to unearth China's and Shanghai's engagement with the Expo. Hosting the Expo was framed as a symbol of China's re-emergence as a global power. Repeatedly emphasized in newspapers was the symbolic meaning of China as a developing country in fighting in a game exclusive to powerful nations in advanced economies (e.g. Pan *et al.*, 2002). China's entry to this game is convincing evidence of the party's accomplishment in materializing the nation's century-long dream (e.g. Pan *et al.*, 2002). Suffice it to say, the symbolic value of the Expo, to avenge the humiliation and to recover *mianzi* (face), transcends the possible and actual economic opportunities. It is no surprise that the Chinese national leaders promised a most successful, most wonderful, most memorable Expo through full-scale institutional support (diplomatic, financial, regulatory, etc.) for Shanghai to the Expo community during the bidding process (Chen, 2010).

For those displaced, such discursive construction created formidable challenges to lodge their complaints or to organize effective protests. Those displaced were not only effectively excluded from legitimate scientific language to speak about the Expo, but were also deprived of the political opportunity by the nationalistic discourses. In defending their right to stay, they would be under the moral pressure of acting against the interest of the national state, meaning the possibility of losing sympathy from the rest of society. Consequently, they had no other choice but to move out. As a matter of fact, moving out was made an obligation through regulations specifically regarding the Expo-induced displacement and dispossession, as discussed below.

Hierarchized disposable lives

Those displaced are certainly not a united group. There are many cleavages within them along the lines of class, gender or generation, which undermined the collectivity for taking control of neighbourhood changes. Here, I follow Butler and Athanasiou (2013) and examine the political construction of uncountability and disposability. I focus on dispossessory arrangements and institutional configurations, which hierarchized the bodies of those displaced and prevented collective actions.

The institutional set-up of the Expo induced displacement and resettlement compartmented those displaced by jurisdictional boundaries. The Expo Park and two major resettlement housing sites occupy land in jurisdictions of four district governments. The Shanghai government established the Expo Land Bank as the legal institution coordinating displacement and dispossession. This land bank negotiated with and entrusted district governments to carry out displacement and dispossession within their own jurisdictions according to the progress of the Expo project. As the de facto displacers, the district governments drew up the most fine-grained regulations on resettlement arrangements on the basis of existing legal regulations. Similar in format, these regulations however vary in the arrangements regarding the overall process of resettlement, the formulas of compensations, the openings in participation and agencies for appeals. Two critical issues were in their discretion – the minimum floor size of resettlement housing for those displaced and minimum monetary compensation per square metre. These technical differences, although not entirely intentional, reinforced the geographical boundaries and created practical difficulties for those displaced to arrive at any common vision regarding the target of their defiance.

In the Chinese context, displacement and dispossession is narrowed to a matter about tangible homes to be demolished. While neighbourhood environment and informal supportive networks might be essential for many residents to get by, the state does not recognize claims for the shared space and sustaining environment. As the state aims to create a legal façade for a violent undertaking, negotiation for resettlement deals is steered around property. As Verdery pointed out, property "specifies what things have what values and who counts as a person" (2003, 18). One of the functions of property rights is to create boundaries, "concerning what belongs to whom and who belongs to or has affinities with some large entity" (ibid.). Therefore, to speak about property in displacement and dispossession is already to individualize and divide.

The property regime divided those displaced into at least three groups, subjecting them to different arrangements and thereby adjusting their expectations and claims in displacement and dispossession. The first group is villagers who lost their homes to the resettlement housing projects. Compared to urbanites who could not ground their collective interest in the land of their communities, villagers could make a stronger claim based on their collective ownership of land. As per law, expropriation of properties in rural and urban areas must follow different procedures. Villagers are entitled to compensation for their lost properties, social security contributions and agricultural outputs. Since land value is excluded in compensation, rural residents are only entitled to residual values of their properties. Even though they might have invested a great deal in purchasing building materials or decorating their houses, the self-built nature of their houses meant that their houses had low economic values in the eyes of the valuators. More importantly, the thorny issues of employment and alternative livelihood were excluded by the authorities, due partially to the liberalization of the labour market and the decline of state-controlled collective industries. While this was less a problem for villagers who had already been employed as industrial labour, especially for the younger generations, it presented challenges for villagers who relied on farming to supplement their incomes.

Ironically, despite their sacrifice being indispensable to the success of land clearance and resettlement, villagers were not officially counted as displaced as a result of hosting the Expo. They were never mentioned in the Expo plan, as if expropriation of their land was taken for granted, in no need of professional labour to justify it. One common complaint from the villagers was that they did not receive similar gestural gifts awarded to those displaced from the Expo Park. They complained not because these gifts were significant in economic value but that these gifts represented state recognition for the sacrifices that the displaced made to the "greater good". As a villager commented on the differential treatment, "they [the urbanites from the Expo Park] are the birth children of the state; we [rural households] were adopted. Adopted children cannot be compared to the birth children" (interview, March 2012).

The second group are tenants. The Expo site provided a foothold for many migrant workers from outside Shanghai due to cheap rents. Unfortunately, they were not compensated during displacement because of the lack of any legally recognized interests. Unlike the complete exclusion of migrant workers, tenants of public or *danwei* (work unit) housing were compensated for their housing rights. This may be partially because these tenants acquired housing as part of their social welfare during the planned economy period. Their status as socialist workers would offer them the moral resources to frame unfair treatment as a further betrayal of the socialist contract. This would potentially undermine the legitimacy of the state. Such claims would carry more weight in the case of the Expo, which attracted global attention.

The arrangement allowed many tenants to purchase resettlement apartments and allow them to become private property owners. However, some tenants of danwei housing had to craft their relations with their danwei in order to get a fair compensation. In the case of the steel company, which occupied nearly half of the Expo site, they were also involved in resettling their workers who could not afford to buy a property in Shanghai and lived in the company dormitories. These workers were organizationally dependent on the company for their employment, residence registration and space for living. Such dependence was abused by their company. The company sent the line managers of workers to negotiate and threatened to dismiss workers if they held out (interview, June 2012).

The third group are property owners from the Expo site. Like villagers from resettlement sites, they had to accept the decision and move out although they were lawful owners. Unlike villagers, they were more ambivalent towards displacement and resettlement. A main reason was the distressed housing conditions of many of those displaced, who did not have the capacity to renovate on their own, let alone purchase a decent apartment from the market. Given the international attention to the Expo and the Chinese state's commitment to delivering a successful Expo, the authorities offered a much higher compensation to them (between 10,300 RMB/m^2 and 13,000 RMB/m^2 depending on locations and timing of displacement) than the normal standards for public projects. However, this was less about caring for those displaced than about securing their cooperation to evacuate from the site in time. Moreover, the authorities adopted a differentiated valuation method to further compartmentalize the homeowners. Self-built housing was valued by referring to the value of public housing instead of commercial housing in planned communities in similar locations. This positions owners of self-built housing in an inferior position in the hierarchy of homeownership, even though they probably had endured more hardships in building their homes over the years through their own investments.

The institutionalized hierarchization did more than dividing those displaced, destroying their social bonds and preventing a political coalition. Worse, they restricted political imagination, turning the "privileged" compensation standard applied to one group or in one locality as a goal of protesting actions from those displaced who fell under lower compensation standards. Their view, hardly self-consciously formed though, is rooted in Confucius ethics that sustains hierarchical social order in social space and posits justice not as measured against universal value but as a relative perception derived from comparison to others in the same social and spatial category. The repressive power of such moral cultures lies in that it restricts, undermines and tames political actions. Distributive justice for them is sought not by overthrowing the oppressive structure or the powerful growth machine in a truly privileged position through collective efforts, but to make an unjust situation less unjust, or as unjust as those in better positions, competing to become less miserable victims. By fostering or tolerating competitive victimhood, the local states in Shanghai were able to clear the land within a short span of two years with few organized resistances.

Conclusion

In this chapter I have discussed the production of new urban spaces in Shanghai through hosting the Expo. By identifying with state-led nationalist discourses and rationalizing the choice of the Expo through technocratic language, the local government was able to carry out a wholesale demolition and force original land users and property owners to move out, without consulting them or seeking their consent for changing land uses. With no opportunities to stay put, those displaced had no choice but to negotiate for better resettlement deals. Yet, the negotiation was narrowed to bargaining for lost properties without taking into account the wider environment and the social networks, from which those displaced derived informal social support. Worse still, the institutional arrangements and regulatory frames created many internal divisions and hierarchies within the displaced, shaping their imagination of political possibilities and fostering competitive victimhood. As a result, a project involving a large-scale property dispossession and displacement was met with little organized, collective resistance. This made it possible for the local state to rationalize urban spatial structure and population distribution and to finance such an ambitious project by pushing up development density in gentrified inner city areas within and surrounding the Expo site as well as in urban fringes. The real estate development agenda and inadequate compensation to original users thus challenges the rhetoric of public interest that justifies the use of compulsory purchase to pursue densification in the case of the Expo.

Of course, densification does not always require compulsory purchase. For instance, denser development on collectively owned rural land can be pursued by rural residents from that collective to accommodate changes in household composition through seeking administrative approval from higher level authorities without invoking compulsory acquisitions. But in most cases, for densification to achieve wider impacts, it requires a certain scale and hence is likely to disrupt existing power relations and vested interests surrounding land uses. In these cases, compulsory purchase is inevitable. The case of the Expo nicely illustrates the politics and the operation of compulsory purchase in contemporary China. Unlike compulsory property acquisitions in socialist times, the construction of institutional foundations for a market economy and the concern for durable dominance of the party-state has brought compulsory purchase under greater legal scrutiny. As a double-edged sword, the law provides protection to victims of compulsory acquisitions by laying out the threshold and procedures of forced evictions, the standards of compensation and available legal remedies, which to some extent restrains the grabbing hands of state-led growth machine. Yet, this protection remains minimal, weak and ineffective due to the political priority to economic growth over legal and social justice. Given the heavy reliance of municipal finance on land revenues, it is in the interest of the local governments to maximize the gap between compensation to original users and returns from land development. This forces local governments to make a calculated trade-off between for-public planning interventions and for-private real estate development through large inflows of capital investment.

Over 50 years ago, Jane Jacobs warned against cataclysmic money in financing drastic transformations and damaging urban fabrics. She urged that "the forms in which money is used must be converted to instruments of regeneration – from instruments buying violent cataclysms to instruments buying continual, gradual, complex, and gentler change" (Jacobs, 1961, 331). Her insight is ever more relevant in the current juncture when cities and the built environment have been increasingly treated as financial assets, and redevelopment of the built environment is being dominated by the logic of finance, which leads to decontextualization of the land uses from local needs (Savini and Aalbers, 2016). This is also what we need to be wary of when conceiving and implementing planned changes in Chinese cities. Wholesale demolition and redevelopment through cataclysmic investment has misplaced the focus to be on city building and land development instead of on the people. Until we start to value more people's needs and facilitate more gradual changes with greater care, planned changes are not only vulnerable to failures but also likely to cause much greater harm. This of course would require simultaneous reforms in public finance to reduce the reliance of municipal finance on land.

References

Abramson D B (2006) Urban planning in China – continuity and change, *Journal of the American Planning Association*, 72(2), 197–215.

Butler J and Athanasiou A (2013) *Dispossession: The Performative in the Political*. Cambridge: Polity Press.

Chen Y (2011) *Shanghai shibo yingxiang jiyi [Memory of the Expo in Photos: a Photographer Witness Lived in Former Site of the Expo Park]*. Shanghai: Shanghai Fine Art Publishing House.

Chen Z (2010) *Shenbo Jiyi: Dashijian Xiaoxijie [Memoir of Bidding for the Expo: Mega-event and Trivial Details]*. Shanghai: Oriental Publishing Center, Shanghai.

Contemporary Shanghai Institute (2003) *Shibohui yu shanghai xinyilun fazhan dataolun jishi [The Expo and the New Round of Development of Shanghai]*. Shanghai: Shanghai People's Press.

Ding C and Lichtenberg E (2011) Land and urban economic growth in China, *Journal of Regional Science*, 51(2), 299–317.

Editorial Committee of Shanghai Expo Planning (2010) *Planning of EXPO 2010 Shanghai China*. Shanghai: Shanghai Press of Science and Technology.

Fewsmith J (2013) *The Logic and Limits of Political Reform in China*. Cambridge: Cambridge University Press.

Flyvbjerg B (1998) *Rationality and Power: Democracy in Practice*. Chicago: University of Chicago Press.

Ho C Y (2013) Bargaining demolition in China: A practice of distrust, *Critique of Anthropology*, 33(4), 412–428.

Ho P (2001) Who owns China's land? Policies, property rights and deliberate institutional ambiguity, *China Quarterly*, 166 (June), 394–421.

Horne J (2007) The four 'knowns' of sports mega-events, *Leisure Studies*, 26(1), 81–96.

Hsing Y-T (2010) *The Great Urban Transformation: Politics of Land and Property in China*. New York: Oxford University Press.

Jacobs J (1961) *The Death and Life of Great American Cities*. Harmondsworth: Penguin.

Lieberthal K (1995) *Governing China: From Revolution through Reform*. New York: W.W. Norton.

Liu X (2009) *The Mirage of China: Anti-Humanism, Narcissism, and Corporeality of the Contemporary World*. New York: Berghahn Books.

Liu Y (2016) From victims to winners: Strategies of negotiation among relocated residents in demolition-compensation events in local China, *China Journal of Social Work*, 9(1), 75–86.

Lubman S B (1999) *Bird in a Cage: Legal Reform in China after Mao*. Stanford, CA: Stanford University Press.

Ong L H (2014) State-led urbanization in China: Skyscrapers, land revenue and "concentrated villages", *China Quarterly*, 217 (March), 162–179.

Pan T S and Liu Z J (2011) Place matters, *Chinese Sociology and Anthropology*, 43(4), 52–73.

Pan Y, Sun W and Chen C. (2002) Yiwan huaxia ernv de qipan, *Wenhui News*, 12 March.

People's Republic of China (2004) *Land Management Act*.

People's Republic of China (2007) *Property Law*.

Poppe W and Young D (2015) The politics of place: Place-making versus densification in Toronto's tower neighbourhoods, *International Journal of Urban and Regional Research*, 39(3), 613–621.

Qu J (2012) Project system: A new system of state governance, *Social Sciences in China*, 33(4), 28–47.

Ren X (2014) The political economy of urban ruins: Redeveloping Shanghai, *International Journal of Urban and Regional Research*, 38(3), 1081–1091.

Rossi U (2013) On the varying ontologies of capitalism: Embeddedness, dispossession, subsumption, *Progress in Human Geography*, 37(3), 348–365.

Savini F and Aalbers M B (2016) The de-contextualisation of land use planning through financialisation: Urban redevelopment in Milan, *European Urban and Regional Studies*, 23(4), 878–894.

Searle G and Filion P (2011) Planning context and urban intensification outcomes: Sydney versus Toronto, *Urban Studies*, 48(7), 1419–1438.

Shao Q (2013) *Shanghai Gone: Domicide and Defiance in a Chinese Megacity*. Lanham, MD: Rowman & Littlefield Publishers, Inc.

Shih M (2010) Governing through law: Rights based conflicts and property development in Shanghai, *Urban Geography*, 31(7), 973–987.

State Council (2011) Regulation on the expropriation of buildings on state-owned land and compensation.

Tan M (2014) Strict control and incentive: Evolution of the policy of linking decrease of arable land with increase of construction land and its local implementation, *China Social Sciences*, 7, 125–142.

United Nations (2016) *Sustainable Development Goals* [Online]. Available at: www.un.org/sustainabledevelopment/cities/ Accessed 25 August 2017.

Verdery K (2003) *The Vanishing Hectare: Property and Value in Postsocialist Transylvania*. Ithaca, NY: Cornell University Press.

Wang H, Tao R, Wang L L and Su F B (2010) Farmland preservation and land development rights trading in Zhejiang, China, *Habitat International*, 34(4), 454–463.

Webber M (2008) The places of primitive accumulation in rural China, *Economic Geography*, 84(4), 395–421.

Wu F (2015) *Planning for Growth: Urban and Regional Planning in China*. London: Routledge.

Wu F (2016) State dominance in urban redevelopment, *Urban Affairs Review*, 52(5), 631–658.

9 Rural densification under China's Link Policy

Long Cheng

Introduction

China has experienced rapid development and urbanisation since the launch of economic reform and openness policy in the late 1970s (Xu *et al.*, 2011). Urban population in China has been increasing by 13–15 million per year since 1978 with the urbanisation rate increasing from 17.9 per cent in 1978 to 54.8 per cent in 2014 (World Bank 2014, 26–41). Cities (including prefecture level and county level cities) sprouted out dramatically from only 193 in 1978 to 657 in 2016, which induced urban built-up areas to expand continuously by 2380 square kilometres per year on average from 1996 to 2012 (National Bureau of Statistics of China 1978; The State Council of the People's Republic of China 2015; Ministry of Housing and Urban-Rural Development of the Peoples's Republic of China 2016). However, rapid urban development also brought about the encroachment of rural land and the removal of villages (Xu *et al.*, 2011). For instance, the amount of cultivated land decreased from 138.39 mha (million ha) in 1978 to 122.07 mha in 2005 (National Bureau of Statistics of China, 2007). This chapter explores how China's Link Policy has addressed this encroachment of rural land through densification of rural housing in China's rapid urbanisation process.

China's unprecedented urban development can be attributed to its two institutional designs: the land use rights (LURs) system and land acquisition (Ding, 2007). LURs were first introduced in the late 1980s when the Bureau of Land Administration was established and the Land Administration Law passed in 1986. The separation of land use rights and land ownership not only improved land use efficiency and enhanced land management, but also generated the land markets in China (Ding, 2003). Two levels of land markets are incorporated into the LURs system: the primary market and the secondary land market. The primary land market refers to the leasing of state-owned land in the cities and towns by local government to users. The users could sell, rent, mortgage and sublease to third parties which constitute the secondary land markets (Chen *et al.*, 2015; Ding, 2007). However, the land market in contemporary China is only used for the land use rights transaction of state-owned land in urban areas. Rural land in China is owned by the socialist rural collectives which are prohibited from selling land to other users (Xie *et al.*, 2002). The only channel where rural land could go into the land market is through the other institutional framework – rural land acquisition. Rural land acquisition is the only type of land ownership transaction in China. As outlined in the regulations of Land Administration Law, in case it's necessary to increase construction land for urban development, the local government must expropriate the rural collective-owned land into state-owned land, and then transfer it to the land users through the primary land market. Therefore, land supply on the primary land market is obviously monopolised by the government (Zhang, 2014). The dominant role of local government in rural land acquisition and land use rights transactions leads to the absence of full market-based compensation for rural land. Thus, the local government could obtain a large amount of land revenue for urban development by means of expropriating the rural land at low cost while selling it to developers at a much higher market price (Ding, 2007).

Potential land revenue from land use rights transfer has become a strong incentive for local government to expand the urban area by expropriating rural land (Zheng *et al.*, 2014). Between 1996 and 2006, China lost about 20 million acres of farmland and the grain output dropped from 508 million tonnes in 1999 to 431 million tons in 2003 (Looney, 2012). Food security has been the main concern of the Chinese government. Confronted with the dilemma of providing land for urban development whilst protecting farmland for agricultural production, a state-led land use policy was proposed in 2005, which was termed "Linking the increase in urban construction land with the decrease in rural construction land" (the Link Policy). Under this scheme, the government could resettle farmers who lived in sparsely located and poorly provisioned farmhouses into concentrated modern communities and then convert their homestead land into farmland. Consequently, more land quotas could be provided for urban construction with the total amount of farmland remaining unchanged. Featuring the densification of rural settlement and compulsory land acquisition in the rural and peri-urban area, the implementation of the Link Policy has been regarded by many local governments as a successful model to improve rural land use efficiency and expedite contemporary China's urbanisation process (Long *et al.*, 2009). However, a review of the latest studies on China's urbanisation particularly regarding rural land acquisition and farmers' resettlement, indicates limited attention has been paid to the impacts of the Link Policy on rural life and residents' satisfaction, especially in English-language research articles.

Given that, this chapter explores the essence of the Link Policy by reviewing policy evolution in China and investigates farmers' perceptions about the policy's impacts on rural life. Empirical research was conducted in Ezhou City, which is the first pilot area implementing the Link Policy in Hubei Province in central China after 2008. One hundred and sixty semi-structured interviews were undertaken by visiting eight concentrated resettlement communities in March 2016. The rest of this chapter is structured as follows. The following section introduces the background of the Link Policy. Then the Policy's implementation in Ezhou is analysed to understand the essence of policy implementation. Key findings from the interviews of resettled farmers are presented afterwards to analyse the impacts of the Link Policy on rural life and understand farmers' satisfaction with these changes. Thereafter, we discuss the issues emerging from the policy implementation in Ezhou with corresponding policy recommendations. The chapter sets out conclusions in the last section.

Rural transformation in reform era: concentrated resettlement

"Building a new socialist countryside" in China

China's economic reform and openness since 1978 have led its market-oriented transition and integration into the global economy through large foreign capital investment (Zhang, 2002). For the three decades of reform, China's economy has been growing at a rate of nearly 10 per cent annually, the highest in the world. This economic achievement also accelerated China's industrialisation and urbanisation process, in which agriculture and countryside have made a big contribution to the development of industries and the cities in China (Long *et al.*, 2009; Long *et al.*, 2011). For instance, besides the decrease of farmland induced by urban development, a series of urban-biased policies resulted in less financial support for agricultural production and improvement of farmers' living conditions. According to data from China's National Bureau of Statistics, the income gap between urban and rural residents widened from 2.57:1 in 1978 to 3.22:1 in 2005 (National Bureau of Statistics of China, 2007). The urban-rural discrepancies have been prominent which hampered China's further social economic development and caused China to deviate from achieving the goal of "making a harmonious society" proposed by the Communist Party of China (CPC) in in its Fourth Plenary Session of the 16th Central Committee in 2004 (Xu and Tan, 2001).

Aiming to address this unbalanced urban-rural development, the central government of China proposed a long-term development strategy termed "Building a New Socialist Countryside (BNSC)" in the 11th Five-Year Plan (2006–2010) enacted in the CPC's Fifth Plenary Session of the 16th Central Committee in 2005. BNSC is a comprehensive rural modernisation strategy that requires "getting industry to support agriculture and cities to support the countryside". It aims to achieve "advanced production, improved livelihood, clean and tidy villages, a civilised social atmosphere and efficient management" in rural area (The State Council of the People's Republic of China, 2006). The first and foremost objective of BNSC is to upgrade rural infrastructure and farmers' living conditions. For a long period of time, most villages in China were described by media as "dirty, disorganised and poor" due to the absence of efficient land use plans and poorly provisioned infrastructures. Farmers prefer building more spacious houses on their allocated homesteads which are randomly scattered in rural areas. As these houses are built on their own without a unified plan or design, there is no disposal or treatment system for sewage, organic garbage and excrement pollutants from livestock and poultry (Long *et al.*, 2009). By contrast, cities could offer better living conditions and job opportunities which attract many rural residents, particularly the young generation, to urban areas. The flow of workforce from rural to urban areas results in the phenomenon of "hollowed villages" where farmers' houses are left vacant either seasonally or permanently (Long *et al.*, 2012). As such, it has been regarded as the fundamental task of BNSC to change the negative impression of current villages and make a prosperous rural life by means of improving the rural residents' living conditions.

An innovative model for BNSC: the Link Policy

Since the adoption of BNSC by the central government, villages have been undergoing a remarkable transformation. New rural communities and facilities have been constructed to improve farmers' living conditions. In response to the decrease of farmland in rural areas, the Chinese central government introduced a strict land quota system to achieve the farmland protection objective which is to maintain the total amount of farmland at no less than 1.8 billion *mu* (120 million ha) in the period to 2020 (Long *et al.*, 2012; Tian *et al.*, 2017). The land quota system, in essence, has been taken as an instrument for government's control of urban land supply (Xiao and Zhao, 2015). Limited land quotas are allocated annually to localities for urban construction following the Land Use Master Plan, which barely meets the local demand for urban expansion.

To provide space for local development under the constraint of the land quota system and objective of farmland preservation, the focus was shifted to rural areas. Targeting the emergence of "hollowed villages" and the long-standing low land use efficiency in rural areas, the Ministry of Land and Resources of the People's Republic of China (PRC) proposed the Link Policy as an innovative land use policy in 2005 under the campaign of BNSC. The Link Policy empowers the local government to relocate farmers into newly built densified communities and reclaim their sparsely distributed homestead holdings for farmland by demolishing the original dwellings. The reduced built-up footprint in rural areas is turned into a new "quota" and transferred to urban areas. Local governments are allowed to trade these quotas with developers and expropriate same-sized farmland at the urban fringe for developers' construction projects. In the process of implementing the Link Policy, the total amount of farmland remains unchanged while more spaces are provided to meet the demand for urban development. Moreover, farmers' living conditions are improved by resettlement into modern communities. Given that, the Link Policy has been adopted widely by local government to coordinate urban and rural development under the strategy of BNSC (Long *et al.*, 2009).

The implementation of the Link Policy in Ezhou

Following the idea of the Link Policy proposed in 2005, the Ministry of Land and Resources designated Hubei, Shandong, Tianjin, Jiangsu and Sichuan as experimental sites for implementation in April 2006. Reflecting on the experiences from the experimental areas, the Ministry of Land and Resources introduced the *Guidelines of the Link Policy Implementation* in June 2008, which is the first official document to detail the working steps of the Link Policy's projects. In November 2008, under the guidance of BNSC, Hubei provincial government approved Ezhou as the first reform pilot city for "Integrated Urban-Rural Development", with the Link Policy being adopted by Ezhou government.

Ezhou is a medium-sized but rapidly growing city in Hubei Province in central China. Geographically, it is located in the middle reaches of the Yangtze River, adjacent to Wuhan, the capital of Hubei Province, with an area of 1500 km^2 and a population of 1.06 million in 2016 (Hubei Statistic Bureau, 2017). In response to the call for integrated urban-rural development and BNSC, the *Ezhou Plan for Urban and Rural Integrated Development* was made in 2008. It aims to build 106 new concentrated modern communities for rural renovation by gradually demolishing and combining the current 320 natural villages (Yang and Xu, 2013). In this process, the Link Policy was a key instrument for coordinating urban and rural development.

In the implementation process of the Link Policy, there are several key elements. First and foremost, it is the platform for property rights transfer. As the rural resettlement and urban development process involves numerous land use rights transfers, the Ezhou government introduced a market mechanism via setting up the Ezhou Rural Property Rights Trading Centre (ERPRTC) in 2012. ERPRTC is affiliated to the Ezhou Bureau of Land Administration and is responsible for land quota collection and transactions between rural localities and urban developers. The second key element is the land quota created and traded within the implementation of the Link Policy. These land quotas are termed "Link Quotas" which are created from rural settlement densification. Rural committees initiate the renovation programme by building densified communities to accommodate resettled farmers. The existing homestead lands are required to be converted into farmland under the supervision of the Ezhou Bureau of Land Administration to make sure the newly created farmland could meet the criteria for cultivation. Link Quota Certificates which label the acreage of these new farmlands are issued to rural committees and all the created Link Quotas go to the ERPRTC for sale to urban developers. Currently in Ezhou, as stated in the *Regulations on Further Developing Rural Land Use Rights Transaction System to Promote Rural Land Consolidation and Rural New Community Construction* enacted by Ezhou municipal government in 2012, a Link Quota has become the prerequisite for any developer who wants to use state-owned land (Ezhou Municipal Government, 2012). This means urban developers are required to hold the same amount of Link Quotas as the state-owned land they intend to develop. These state-owned lands on the land market come from the rural land acquisition. Ezhou Bureau of Land Administration expropriates farmland at the urban fringe by using equivalent Link Quotas collected from rural committee and then leases these lands to developers. The generation and use of Link Quotas is visually shown in Figure 9.1.

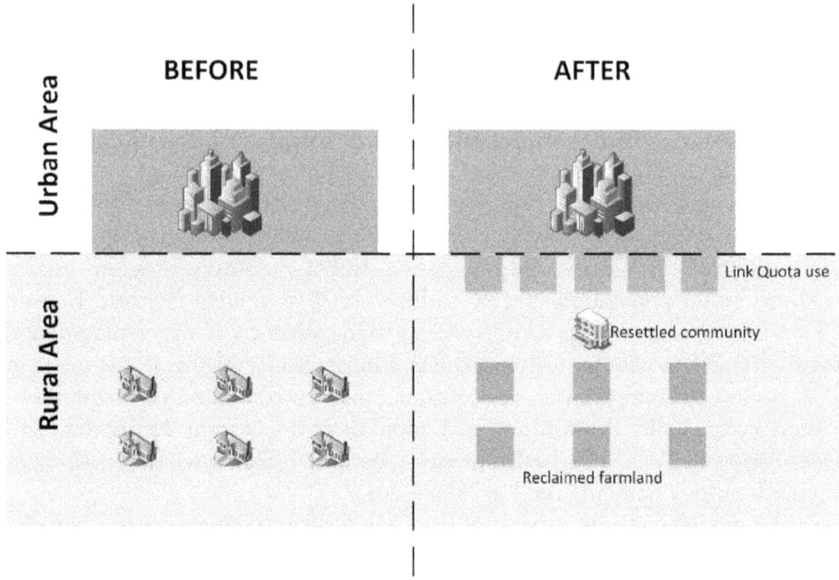

Figure 9.1 Link Quota generation and use.

The price of Link Quotas is currently fixed by ERPRTC at 160,000 CNY/mu (0.37 million USD/ha[1]) which is calculated based on the average cost of rural resettlement and land consolidation. The revenue from Link Quotas excluding Link Quota transaction fees taken by ERPRTC goes to the rural committee for rural land consolidation, new community construction and compensation for resettled farmers. Besides the payment for Link Quotas, developers also have to pay the land premium for using the state-owned land. The price of the land premium for parcels varies with different locations. The Ezhou Bureau of Land Administration normally offers a baseline for the land premium of a given parcel and invites several developers to bid at auction. The final price of the land premium is determined by the highest price offered by the successful competitor. The land premium has become a primary source for local government income in China while only a small proportion is used to compensate land-loss farmers at the urban fringe. Developers who want to use the land for development have to pay for both the Link Quota and land premium. The money and land flows in the implementation process are shown in Figure 9.2.

Policy impacts and farmers' satisfaction

Since the adoption of the Link Policy by Ezhou government in 2009, eight official regulation documents have been announced to regulate the implementation process, and 152 Link Quotas collected from 32 villages have been sold to

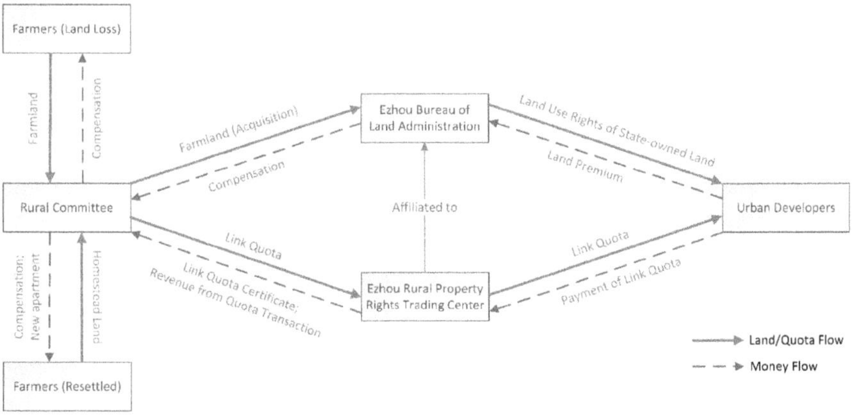

Figure 9.2 Money and land flows in Link Policy implementation.

developers with an area of 4763.46 mu through ERPRTC. To understand the policy impacts on rural life and farmers' satisfaction, I undertook 160 semi-structured interviews of resettled farmers from eight concentrated resettled communities.

The questionnaire consisted of three main parts: (1) farmer's understanding of the Link Policy; (2) changes of farmer's life after resettlement; (3) compensation and farmer's satisfaction with the Link Policy. The demographic features of interviewed farmers are shown in Table 9.1.

Table 9.1 Demographic features of interviewees

Attributes	*Classification*	*Proportion (per cent)*
Gender	Male	57.33
	Female	42.67
Age	<30	5.33
	30–39	5.33
	40–49	30.67
	50–59	26.67
	60 & over	32.00
Education	Primary school or below	35.33
	Junior middle school	40.00
	Senior high school	22.67
	Bachelor or above	2.00
CPC Membership	CPC Membership	20.00
	Non-party Membership	80.00

Impacts of the Link Policy on rural life

Drawing on the fieldwork in resettled communities, the impacts of the Link Policy on rural life can be considered from two main aspects: changes in farmers' living conditions and their lifestyle.

In terms of changes in farmers' living conditions, 71.33 per cent (107/150) indicated the new community provided a better place to live than before especially for the improvement of sanitation and infrastructure. The improvement of rural living conditions could also be proved through the author's visits to these communities. As shown in Figure 9.3, compared to farmers' old dwellings, the new communities offer more urbanised apartments with integrated facilities such as the grocery store, sports field, schools, clinic, fitness equipment and shuttle bus to the city. Moreover, running water, gas and electricity are all available in each household. Farmers in the newly built concentrated communities are now enjoying an urbanised life with similar living environment and facilities to their city counterparts without moving away from their origin. The average distance between farmers' old places to the new community is 0.96 km for those interviewed, while farmers' new dwellings are in better locations that are connected with the main route to the city or other towns. As indicated from the

Figure 9.3 Living conditions before and after resettlement.

interviews, the average travelling time from home to city decreased from 0.6 hours to 0.45 hours. However, some issues are also revealed from fieldwork. For instance, 16 per cent (24/150) of interviewees complained about the tiny living room in their apartment compared to their original houses and could not adapt to this "urban" life, especially the old generation. Notably, the average floor area in each household decreased from 232.51 m^2 to 213.21 m^2 after resettlement and farmers lost their courtyard which was normally used for storing farming tools, cultivating vegetables for their own consumption, and raising poultry or livestock.

In regard to the changes of rural lifestyle, corresponding to the access to better facilities in the new communities, living costs also increased after resettlement. Just over 74 per cent (112/150) of interviewees considered food shopping as the main increase in their monthly expenses due to the loss of farmland and courtyard where they used to grow grain and vegetables, and raised poultry for their own consumption. Seventy-one per cent (107/150) of interviewees said that bills for running water, electricity, gas, cable TV and the Internet dominated increases in their living costs. The average monthly expenditure in interviewed households increased from 1273.33 CNY to 1994.67 CNY, in which spending on food shopping increased from 683.87 CNY to 997.80 CNY. Concurrently, the average household annual income also increased dramatically from 40,318.67 CNY to 59,512.93 CNY due to the transition from agricultural production to non-agricultural production. After resettlement, farmers' annual time on agricultural production decreased from 6.22 months to 4.64 months and their contracted farmland allocation from the rural committee also decreased from 5.64 mu to 2.48 mu on average. Nineteen per cent (29/150) of interviewees indicated their current dwellings are far away from their farmland. The average time walking to farmland increased from 11.35 minutes to 18.62 minutes. This contributes to farmers taking on more non-agricultural work after resettlement.

Farmer's satisfaction with the Link Policy

As the first experimental city for the integrated urban and rural development programme, Ezhou's achievements in carrying out the Link Policy were always publicly reported as a successful model for other cities (Yang and Xu, 2013). However, with farmers as the key stakeholders in the implementation, their satisfaction is of vital importance for policy evaluation and adjustment to achieve better implementation. To explore farmers' satisfaction with the Link Policy, interviews were conducted from the perspectives of participation willingness, policy understanding and compensation received.

Overall, 48.67 per cent (73/150) are satisfied with the implementation of the Link Policy. For the reasons, all the respondents indicated the improvement of living conditions, particularly the sanitation conditions. Sixty-two per cent (45/73) are satisfied with the recreational activities in the new communities which enriched their life. Fifty-eight per cent (42/73) of the interviewees referred to the public services such as the clinics in the new communities.

On the contrary, 32.67 per cent (49/150) are not satisfied with the Link Policy programme. Among those, low compensation and the increased living costs are the primary reason for all respondents. In addition, 59 per cent (29/49) complain about the loss of their farmland and the insufficient support or placement for the land loss farmers.

In terms of farmers' willingness to participate in the programme, 74 per cent (111/150) of the interviewees indicated that the rural committee had consulted them on their willingness before initiating the programme: 47 per cent (70/150) were willing to participate, with 45.72 per cent (32/70) having strong willingness due to the attraction of urbanised dwellings in the new communities. However, 29.33 per cent (44/150) were unwilling to move, where the primary reason was disagreement with the compensation being offered. Farmers' knowledge of the Link Policy also has impacts on their satisfaction. From those interviewed, 62.67 per cent (94/150) had heard about the Link Policy while 37.33 per cent had no idea about it. Consultation meetings were the primary channel for those who had been informed. Farmers' knowledge of the Link Policy is still very vague and some may even have misinterpretation about it. For instance, regarding the prime policy objective, 48.33 per cent considered that the policy was an alternative way for government to grab land from farmers, a view that will directly affect their satisfaction with the policy.

Another key component related to farmers' satisfaction is the compensation for resettlement. As mentioned above, all the farmers dissatisfied with the policy selected insufficient compensation as the main reason. Moreover, 60 per cent (90/150) were not satisfied with the current compensation because the money they received could not offset the expense of resettlement. For instance, in the case of Hengshan community, the compensation for a farmer's old house is 80 CNY/m^2. By contrast, the price for the new apartment is 720 CNY/m^2. There are also a variety of extra costs after resettlement such as furnishing the home and shopping for appliances. The cost of resettlement has been a heavy burden for these farmers. As indicated from interviews, 33.33 per cent (50/150) of the interviewed households were in debt because of relocation. The dissatisfaction with compensation is also caused by insufficient farmers' participation. Seventy-two per cent (108/150) of the respondents indicated that no consultation meeting regarding compensation had been organised and that they had never been informed about participating in a consultation discussion.

Discussion

Understanding the essence of the Link Policy

Since the Chinese government introduced the "Build a New Socialist Countryside" (BNSC) strategy in 2005, a variety of rural transformation programmes have taken place in China with the common feature of concentrated resettlement. The Link Policy was proposed to be practiced according to the philosophy of BNSC and has been regarded as an effective approach in achieving the objective of integrated urban-rural development by many local governments. Though it has been implemented for years, there are still some misinterpretations by local government and farmers. From the farmers' perspective, few of them consider the Link Policy as another way of land acquisition while the primary policy objectives are preserving farmland and improving rural living conditions. Moreover, some localities pursue more revenue from Link Quota transactions by initiating the resettlement programme irrationally. Taking Ezhou as an example, since the establishment of ERPRTC in 2012, 152 transactions of Link Quotas have been accomplished with an area of 4763.46 mu which brought revenue of 0.76 billion CNY. However, these Link Quotas sold by the local government were far more than those collected which only aggregated to 1200 mu. The sum of 168.29 million CNY of the quota revenue went to rural collectives and farmers (Chen *et al.*, 2016). ERPRTC regards this pre-sold behaviour as a way of fund raising for implementing the Link Policy though this was warned against by the Ministry of Land Resources in 2012.

As such, understanding the essence of the Link Policy plays a pivotal role in achieving the policy objectives and improving farmers' satisfaction. Drawing from the practice of the Link Policy in Ezhou, it could be regarded as a hybrid of three key policy instruments: land acquisition, settlement densification and Transferable Development Rights (TDR). The Link Quota creation process involves the acquisition of farmers' homestead land, concentrated resettlement and the deprivation of farmers' development rights on their homestead land. This process takes place in areas where farmers' living conditions urgently need to be improved, or in the "hollowed village" where the land use efficiency is very low. Farmers' development rights on homestead land are transferred to the places where urban developers intend to apply the Link Quota. Local government also intervenes in this quota usage process by expropriating farmland on the urban fringe for developers. Though this process refers to the transfer of development rights, the Link Policy is not exactly the same as the TDR model in the United States because of the government involvement. Farmers could not directly trade the Link Quota or the development rights to urban developers, because government dominates the land market and possesses the ownership of these lands. Overall, through the densification of rural settlements and the transfer of Link Quotas from rural areas to urban developers, the Link Policy could retain the farmland unchanged in the rapid urbanisation process and improve farmers' living conditions. However, the government's role in the three key policy instruments of the Link Policy should be constrained and more market mechanisms introduced.

Improving farmer satisfaction with the Link Policy

As articulated in the *Regulations on Implementing the Link Policy* enacted by Ministry of Land Resources in 2008, the implementation process should protect farmers' interests and improve their well-being. However, the interview data indicated that less than half of the respondents are satisfied with the implementation of the Link Policy. As such, improving farmers' satisfaction should be treated as the key objective for policy implementation.

The interviews undertaken in Ezhou identified farmers' participation and compensation as the two main aspects where farmers were not satisfied with the Link Policy. In terms of farmers' participation, there should be more involvement in the implementation process. During the planning stage of the Link Policy, farmers' willingness needs to be considered through consultation to select the villages where Link Quotas are provided. Meanwhile, the cadres in selected villages should work more on the interpretation of the Link Policy for farmers and invite them to consultation meetings on the design of resettled communities to make sure the new places can meet their demands for living and production. Moreover, more than half of the interviewees mentioned they are excluded in the compensation plan-making process and are not satisfied with the current compensation received. As such, it should be guaranteed that farmers fully engage in the discussion about formulating the compensation scheme before the project is implemented so as to improve their satisfaction. The current compensation plan contains two items: compensation for the farmland and compensation for the old house. As mentioned above, compensation for the old house is unable to offset the purchase of a new apartment, not to mention other extra costs caused by resettlement. Farmland has been the principal source of livelihood in China's rural society. Farmers would like to retain their farmland by paying an even higher price than they have received for the loss of farmland. The interviews asked farmers' willingness to pay to keep their farmland: the average price cited was 24,573.33 CNY, while the average current monetary compensation is 17,521.52 CNY. The underlying reason for the farmers' stated price concerns livelihood sustainability. Farmland could provide farmers with a long-term sustainable income for living, whereas the monetary compensation they receive is a lump sum payment which generates farmers' concerns about their future livelihood. As such, local government should support farmers to achieve sustainable livelihoods after resettlement such as arranging job opportunities at local enterprises. Moreover, a larger proportion of Link Quota revenue and land should be allocated to resettled farmers to compensate the sacrifices they have borne in the implementation of the Link Policy.

Conclusion

China's rural society is undergoing an unprecedented transformation under the Link Policy. Densified modern rural communities sprout out while inefficiently used farm homestead land is reclaimed into farmland. The implementation of the Link Policy not only enables farmers to get access to more urbanised living conditions but also offers more land quotas for urban construction without a decrease in farmland. To understand the impacts of the Link Policy on rural life and explore farmers' satisfaction with the policy implementation, this chapter has taken Ezhou as a case study area. By undertaking semi-structured interviews with the resettled farmers in the densified communities, it can be concluded that certain impacts of the Link Policy on rural life have been emerging from the changes in farmers' living conditions and their lifestyles. New communities in villages are more urbanised and integrated with modern infrastructure and facilities. However, the increase in living costs in new places and insufficient compensation for resettlement have resulted in farmers' dissatisfaction with the Link Policy, which could inspire resistance to further implementation and may even cause conflict between farmers and local governments (Long *et al.*, 2012). As such, this chapter suggests that local government should include farmers in the project planning process and allocate a larger proportion of land revenue to compensate resettled farmers. Thus the implementation of the Link Policy would gain more support from farmers and a more coordinated urban-rural development would be achieved in the process of China's urbanisation.

This chapter has shown how the Link Policy addressed the focus of the book in regard to five of its central questions. First, the background of the Link Policy indicates that circumstances justify densification through taking property rights. Facing the limited construction land allocation under the current land quota system and concerns about food security resulting from the loss of farmland, the Link Policy was introduced by the Chinese government. The densification of rural housing through rural property acquisition could provide more land quotas for urban development without the decrease of farmland. Farmers' living conditions are improved due to the densification process, which has been identified from our field observation and interviews. Second, in terms of the role developers play in the compulsory acquisition, we identified that urban developers accelerated the rural property acquisition process by pursuing Link Quotas. Given the strict land quota system in China, the Link Policy has been an effective channel to provide land for urban construction. Through the densification of rural settlement and the reclamation of farmers' original homesteads, extra farmland areas are created, termed "Link Quotas", which could be transferred to developers for their construction projects at the urban fringe. In Ezhou's case, the Link Quota has been the prerequisite for all urban development projects. Developers' urgent demand for Link Quotas accelerates the process of homestead land acquisition and densified resettlement in rural area. Third, regarding the question about compensation for property acquisition, the Link Policy offers a lump sum monetary compensation for farmers' original farmland and dwellings. Densified modern communities are constructed and provided with better living conditions by local government for resettled farmers. Both the monetary compensation for farmers and the cost of resettled communities are financially supported from the land premium and the revenue from Link Quotas that developers have paid. However, drawing on our fieldwork, current compensation can be argued as still not enough to offset farmers' costs of resettlement. Moreover, changes of lifestyles resulting from densification of rural housing have increased farmers' living costs, while a less supportive scheme has been provided for land loss farmers to sustain their livelihood. This has resulted in farmers' dissatisfaction with the policy. Fourth, in regard to the question concerning types or locations of densification, the Link Policy indicates that villages with low land use efficiency and poor living conditions are normally selected as the areas where farmers' dwellings are expropriated through the densification of rural housing. The densified rural communities are in better locations that are connected with the main route to the city or other towns. Shuttle buses are also provided for farmers travelling between towns and homes. Finally, the Link Policy denotes that it is the obligation of rural committee to provide alternative housing in lieu for those resettled farmers. Densified rural communities should be constructed before resettlement is initiated. However, in Ezhou, the farmers have to pay for their new apartments while the compensation they receive from the Link Policy is insufficient for this payment. We suggest that a fairer compensation scheme with more consideration of farmers' needs and more participation by farmers is required to improve farmers' satisfaction with the policy. As such, rural densification under the Link Policy could proceed more successfully with the support of farmers and contribute to the coordination of urban and rural development in China's urbanisation.

Note

1 1 CNY = 0.15 USD.

References

Chen W, Pan R and Xiao H (2016) Ezhou model of land quota transaction under the Link Policy, *China Land*, 2, 40–42.

Chen Z, Wang Q and Huang X (2015) Can land market development suppress illegal land use in China?, *Habitat International*, 49, 403–412.

Ding C (2003) Land policy reform in China: Assessment and prospects, *Land Use Policy*, 20(2), 109–120.

Ding C (2007) Policy and praxis of land acquisition in China, *Land Use Policy*, 24(1), 1–13.

Ezhou Municipal Government (2012) *Regulations on Further Developing Rural Land Use Rights Transaction System to Promote Rural Land Consolidation and Rural New Community Construction*. Ezhou Municipal Government.

Hubei Statistic Bureau (2017) *2016 Ezhou Economic and Social Development Statistics Report*. Accessed 8 May 2016 at www.stats-hb.gov.cn/wzlm/tjgb/ndtjgb/ezs/ezs/108974.htm.

Long H, Li Y, Liu Y, Woods M and Zou J (2012) Accelerated restructuring in rural China fueled by "increasing vs. decreasing balance" land-use policy for dealing with hollowed villages, *Land Use Policy*, 29(1), 11–22.

Long H, Liu Y, Wu X and Dong G (2009) Spatio-temporal dynamic patterns of farmland and rural settlements in Su–Xi–Chang region: Implications for building a new countryside in coastal China, *Land Use Policy*, 26(2), 322–333.

Long H, Zou J, Pykett J and Li Y (2011) Analysis of rural transformation development in China since the turn of the new millennium, *Applied Geography*, 31(3), 1094–1105.

Looney K (2012) *The rural developmental state: Modernization campaigns and peasant politics in China, Taiwan and South Korea*. Doctoral dissertation, Harvard University.

Ministry of Housing and Urban-Rural Development of the People's Republic of China (2016) *China Urban-Rural Construction Statistical Yearbook*. Beijing: China Statistics Press.

National Bureau of Statistics of China (1978) *China Statistical Yearbook*. Beijing: China Statistics Press.

National Bureau of Statistics of China (2007) *China Statistical Yearbook*. Beijing: China Statistics Press.

The State Council of the People's Republic of China (2006) *Guidelines of PRC's 11th Five-Year Plan for National Economic and Social Development*. Beijing: People's Press.

The State Council of the People's Republic of China (2015) *The New Urbanization Plan (2014–2020)*. Accessed 16 March 2014 at www.gov.cn/gongbao/content/2014/content_2644805.htm.

Tian L, Guo X and Yin W (2017) From urban sprawl to land consolidation in suburban Shanghai under the backdrop of increasing versus decreasing balance policy: A perspective of property rights transfer, *Urban Studies*, 54(4), 878–896.

World Bank (2014) People, in *World Development Indicators 2014*. Washington DC: World Bank.

Xiao Y and Zhao J (2015) Fixing China's distorted urban land quota system, *Paulson Policy Memorandum*. Chicago: The Paulson Institute.

Xie Q, Ghanbari Parsa A R and Redding B (2002) The emergence of the urban land market in China: Evolution, structure, constraints and perspectives, *Urban Studies*, 39(8), 1375–1398.

Xu W and Tan K C (2001) Reform and the process of economic restructuring in rural China: A case study of Yuhang, Zhejiang, *Journal of Rural Studies*, 17(2), 165–181.

Xu Y, Tang B and Chan, E H W (2011) State-led land requisition and transformation of rural villages in transitional China, *Habitat International*, 35(1), 57–65.

Yang N and Xu X (2013) Ezhou model for integrating urban and rural development, *Hubei Daily*. Accessed 12 May 2013 at http://hbrb.cnhubei.com/HTML/hbrb/2013 0512/hbrb2039000.html.

Zhang K H (2002) What explains China's rising urbanisation in the reform era?, *Urban Studies*, 39(12), 2301–2315.

Zhang S (2014) Land-centered urban politics in transitional China – can they be explained by Growth Machine Theory?, *Cities*, 41, 179–186.

Zheng H, Wang X and Cao S (2014) The land finance model jeopardizes China's sustainable development, *Habitat International*, 44, 130–136.

10 Land acquisition in Singapore

Taking and giving

Alice Christudason

Background and introduction to land acquisition in Singapore

It has been just over 50 years since Singapore gained independence and became an *ex*-British colony. Comprising only a little more than 700 square kilometres, Singapore today has one of the highest GDP *per capita* in the world. The vast majority of people own and live in affordable, quality public housing, and the city-state is served by a comprehensive rapid transit network (Urban Redevelopment Authority, 2012a).

Singapore's Changi Airport was developed to become the world's most awarded and sixth busiest for international traffic, serving more than 58.7 million passengers annually (Changi Airport Group, 2017). The city buzzes with nightlife and activities, playing host to major international events like the Formula 1 Grand Prix, with easy access to a comprehensive network of parks, greenery and activated waterways, all bustling with biodiversity (Urban Redevelopment Authority, 2013).

It is difficult to believe that not that long before, when Singapore was granted internal self-government in 1959, an estimated 300,000 people were living in overcrowded slums and squatter settlements with no sanitation, utilities or healthcare. Another 250,000 were in rented partitioned cubicles in old shophouses within the city (Centre for Liveable Cities, 2014a). The proliferation of urban slums posed fire hazards and became breeding grounds for disease and crime. The situation was so dire that the British Housing Committee referred to Singapore as 'one of the world's worst slums', a 'disgrace to a civilised community' and a 'chaotic and unwieldy megapolis … [created by] … by haphazard and unplanned growth' (Centre for Liveable Cities, 2015).

With a GNP per capita of less than US$320, Singapore was a third-world nation with poor infrastructure and limited capital. Low-end commerce was the mainstay of the economy with a handful of industries which produced only for domestic consumption. With massive unemployment and labour unrest, job opportunities had to be created through an environment conducive to industrial development.

Yet, Singapore, a society destined to fail in 1965 (Mahbubani, 2015), morphed into one of the world's greatest success stories within just five decades. This chapter charts Singapore's meteoric physical transformation from the perspective of *inter alia*, legislation on compulsory land acquisition ('acquisition') and urban land policy. It considers how the nature and scale of acquisition evolved to achieve socio-economic growth and national development, resulting in a high density built environment manifested by affordable public housing, industries and infrastructure. In this regard, the importance of the forward-looking and no-holds barred approach of Singapore's founding Prime Minister Mr Lee Kuan Yew (Mr Lee) and his team cannot be over-emphasised. Their multi-faceted and deliberate efforts were the drivers for Singapore's physical transformation through land acquisition.

Basic tenets of acquisition

In this chapter, acquisition refers to the forcible taking of land by the State from private individuals for public benefit. The oft-cited justification (Khublall, 1994) for acquisition is *salus populi est suprema lex* or that the welfare of the people or of the public is the paramount law, and *necessitas publica major est quam privata* – that public necessity is greater than private. While it is difficult to disagree with this, there is little consensus on the nature of protection for affected landowners. Many of the key decisions pertaining to urban development in Singapore at the time were taken on the advice of Dutch economist, Dr Albert Winsemius. He was appointed by Mr Lee as chief economic advisor to the Singapore Government. He held the position from 1961 to 1983, long enough to have a real impact on Singapore's development. Dr Winsemius' advice led to the establishment of the major industrial estate at Jurong, and emphasis on the provision of basic infrastructure including airports and seaports to support these industries (Ong, 2016).

Approach of chapter

The next section commences with Singapore's historical background as a fledgling nation with particularly scarce land resources. It sets out the circumstances which led to legislation which bestowed the State with wide-ranging acquisition powers, and its ramifications. Acquisition provisions were closely matched by *compensation* provisions which were meticulously crafted to suppress compensation, as it was paid out of public coffers.

The basic tenets of acquisition underpinned the compensation provisions. In addition, the compensation provisions reflected the principles of redistributive justice – that private individuals should not enjoy windfalls from public expenditure, and that compensation costs should not unduly burden the city-state. This is dealt with later in the chapter.

Historical background and acquisition measures pre-1966

Sir Stamford Raffles, on behalf of the British East India Company, founded Singapore in 1819. By a Treaty in 1824, Singapore was formally ceded to the British

Government whose general land policy was to create more revenue by granting to the subjects mostly freehold and leasehold titles. In the later years, the Colony reserved the right to resume possession of land for essential public developments.

After 1854, most leasehold titles included covenants which empowered the Crown to resume possession of the land for infrastructure in consideration of compensation.

In the early twentieth century, rapid population growth and increased commercial activities necessitated a better system of administration. The Land Acquisition Ordinance in 1920 was enacted, and conferred upon the Governor a general power to acquire private land for public purposes. The principles underlying the Ordinance were drawn heavily from the Indian Land Acquisition Act of 1894 (Lim, 1968).

Singapore pre-independence: key problems and obstacles to acquisition

The People's Action Party (PAP), Singapore's first elected government under Mr Lee's leadership, took office in 1959 after gaining independence from the British. At the time, fewer than 10 per cent of people in Singapore owned a home. The housing shortfall was about 14,000 units per year and the government had to meet the needs of a population that increased at the rate of 4.5 per cent between 1947 and 1957 (Loh, 2008). The PAP committed to the urgent task of ramping up the supply of public housing. However, the task was daunting, with squatters' colonies dotting the landscape and lining the major trunk roads. In the post-war baby boom, there were often six to seven children per family. Housing supply needed to grow 20 times faster in order to meet demand (Urban Redevelopment Authority, 2013). Also, more opportunities for employment had to be created for the survival of the new city-state.

While ambitious and massive projects had to be undertaken for these purposes, the severe lack of publicly owned land posed a serious impediment to redevelopment plans. The existing legal framework was woefully inadequate to enable the Government to swiftly develop public housing and promote industrialisation at low cost. At the heart of Parliamentary debates on acquisition was, what was the best way to balance national development needs with the right to private property and adequate compensation (Centre for Liveable Cities, 2014a)?

Constitutional safeguard of property rights

While Singapore was part of the Federation of Malaya between 1963 and 1965, the Federation's Constitutional safeguard of property rights applied to Singapore. However, when Singapore seceded from Malaya and gained sovereignty in 1965, the safeguard was excluded from Singapore's new Constitution. The omission was described as deliberate (Ministry of Culture, 1966): '[Otherwise] land reforms would be severely hindered … even impossible as it would have opened the door for litigation and adjudication by the court regarding what was adequate compensation.' The Bill for the Constitution was introduced in 1964; this was the first in a series of well-executed steps to implement focused and result-oriented legislation on acquisition in Singapore.

Post-1966 acquisition policies

Post-independence, most Singaporeans were immigrants. Not unexpectedly, the citizenry did not feel a natural or strong sense of belonging to their newly independent country. Thus, building up a sense of nationhood became a priority for the newly elected Government. With land being scarce, in the face of poor living conditions and the Government's limited funds, there was an urgent and pressing need to acquire an adequate supply of land at a manageable cost (Centre for Liveable Cities, 2014a).

A home-owning society

The legislative mechanism devised to achieve the above was the Land Acquisition Act ('the Act') 1966.[1] Mr Lee's vision of providing universal home ownership and a quality living environment pervaded the acquisition efforts. He said (Lee, 2000, 116–117):

> My primary preoccupation was to give every citizen a stake in the country and its future. I wanted a home-owning society. I had seen the contrast between the blocks of low-cost rental apartments, badly misused and poorly maintained, and those of house-proud owners, and was convinced that if every family owned its home, the country would be more stable … I had seen how voters in capital cities always tended to vote against the government of the day and was determined that our householders should become homeowners; otherwise we would not have political stability. My other important motive was to give all parents whose sons would have to do national service,[2] a stake in the Singapore their sons had to defend. If the soldier's family did not own their home, he would soon conclude he would be fighting to protect the properties of the wealthy. I believed this sense of ownership was vital for our new society which had no deep roots in a common historical experience.

Scope of acquisition provisions under the Act

Section 5 empowers the State, through the Singapore Land Authority ('SLA') to acquire land compulsorily. It states that whenever any particular land is needed:

a for any public purpose;
b by any person, corporation or statutory board, for any work or an undertaking which, in the opinion of the Minister, is of public benefit or of public utility or in the public interest; or

c for any residential, commercial or industrial purposes, the President may, by notification published in the *Gazette*, declare the land to be required for the purpose specified in the notification.

'Public purpose'

What constitutes 'public purpose, benefit, interest and utility' is not defined. It appears that the 'public' does not have to embody the entire community; a section of the community will suffice (Khublall, 1994, 40). In *Hamabai Framjee Petit v Secretary of State*,[3] the Privy Council accepted that public purpose 'must include a purpose, an object or aim, in which the general interest of the community, as opposed to the particular interest of individuals, is directly and vitally concerned'.

Public purpose and comprehensive redevelopment

Section 5(1)(a) and (b) both contain the term 'public' while, significantly, clause (c) does not. The omission was justified as follows (Singapore Parliament, 1966, C31):

> What we have in mind is the acquisition of land for implementation of urban renewal plans which may at some future envisage the acquisition of large pieces of encumbered or occupied lands which, though ready for development, cannot be developed by the owners themselves because of various legal and occupational hindrances.
>
> The Government may acquire such lands and subsequently ... alienate them to private individuals for development in accordance with an overall urban renewal plan.

Acquisition is usually the only practical way forward to achieve site assembly for comprehensive redevelopment. This has been the case in Hong Kong (Planning, Environment and Lands Branch, 1996) and also in Singapore's context (Hui *et al.*, 2004). Thus Section 5 provides for the acquisition of land for development not only by the Government, but envisages that that the private sector may be roped in to engage in the densification effort during comprehensive redevelopment. Comprehensive redevelopment usually involves the amalgamation of land parcels to achieve redevelopment which is more thorough, rather than piecemeal. A unified plan can design for complementarities. For example, with slum clearance, housing can be linked with local shops, schools and open spaces with accessibility to shopping centres and places of work.

The widely phrased section 5(3) facilitated the freeing up of privately owned prime land. The planning authority could avail itself of many opportunities to redevelop Singapore's city-centre. A more recent example is the redevelopment of the Outram area. To achieve comprehensive redevelopment, Pearls Centre will be acquired and amalgamated with adjacent land to allow for better land optimisation around the new Mass Rapid Transport (MRT) station for future densification (Sim, 2012).

Land value capture

Public investment in say, rail and roads generates huge increases in surrounding land values. In part, the increase derives from improved accessibility for existing residents and businesses. Windfalls also arise when land has been rezoned to capitalise on higher development opportunities generated by the new infrastructure. However, ideally, public investments should maximise community benefit. It is therefore unfair and inefficient when private landowners incidentally enjoy windfalls from public investments.

The general consensus is that there should be a recapture through betterment levies, of some of the increased business revenue or land value, that is, the 'unearned increment'. It can then be clawed back and allocated towards the initial costs of infrastructure provision (Gurran and Lawler, 2016). Much of the infrastructure central to Singapore's successful transformation has been funded through land value capture. This is operationalised through sale of land to private developers, and taxation of development uplift (Loo, 2017) through betterment levies in the form of development charge and differential premium.[4] These are imposed when planning permission is granted to carry out development on a site for a more valuable zoning use, or in excess of the existing plot ratio, i.e. higher density.

Public purpose – the 'conclusive evidence' provision

By section 5(3), the President's declaration that land is required for the purpose specified in the Government *Gazette* is conclusive. In *Galstaun & Anor v Attorney General*,[5] Chua J. said that:

> The Government is the proper authority for deciding what a public purpose is. When the Government declares that a certain purpose is a public purpose it must be presumed that the Government is in possession of facts which induce the Government to declare that the purpose is a public purpose.

In light of the above, is it even within a court's jurisdiction to question the basis of the exercise of ministerial discretion on the acquisition? This issue is addressed in the section below on challenges to acquisition. But first, the procedure for acquisition is discussed.

Acquisition procedure

Following the President's declaration, the Collector of Land Revenue ('the Collector') implements acquisition proceedings; these include issuing the notification, the Collector's Inquiry and the taking of possession. The process is depicted in Figure 10.1.

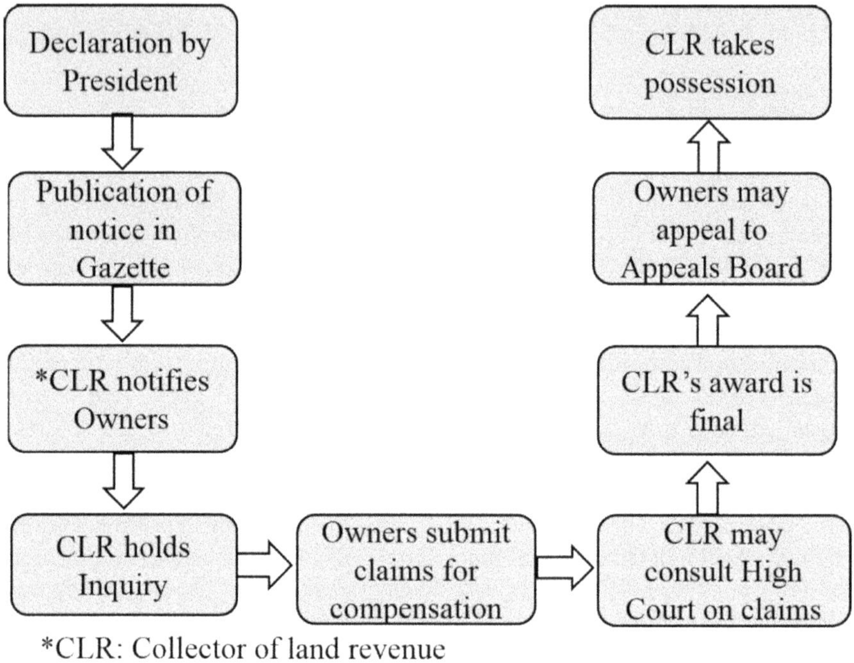

Figure 10.1 Land acquisition procedure.

Source: author's adaptation from Part II of the Act.

Challenges against acquisition

Even in jurisdictions where there is Constitutional protection of property rights, acquisition powers have been described as 'draconian'[6] and a 'serious invasion of private proprietary rights'.[7]

To redress this, in most common law jurisdictions, unlike in Singapore, an aggrieved landowner may have a *statutory* right to object to the acquisition.

In light of the trilogy of lack of Constitutional safeguard for property rights, the wide-ranging powers conferred by section 5, and the conclusive evidence provision, the above descriptions are *a fortiori* in Singapore's context. It is therefore incumbent on the Courts to ensure that an acquiring authority acts within its statutory powers and in good faith. In this regard, the doctrine of *ultra vires* and bad faith may be relied on to declare the acquisition unlawful by way of *certiorari*.[8] Several cases which relied on this doctrine at the Court of Appeal are considered below.

Bad faith and ultra vires

In *Basco Enterprise Pte Ltd v Soh Siang Wai,*[9] property which was acquired for urban redevelopment was used for preservation or conservation. The owner alleged that this was *ultra vires*. The Court held that it is 'arguable that conservation comes within the definition of urban redevelopment'. This is despite the common understanding that redevelopment generally involves demolition and rebuilding (Khublall, 1994, 31).

An exceptionally interesting case on acquisition involving three religious institutions is *Eng Foong Ho & Others v Attorney General*.[10] A Buddhist temple, a Hindu Mission and a church were all religious institutions located within the same area. To make way for an MRT line, the Government acquired only the temple. Temple devotees argued that their Constitutional right to freedom of religion was violated. However the Court found that the Collector's decision to acquire the temple property was based *solely* on planning considerations. In the absence of an allegation of arbitrary action by the Government, the Court decided that the acquisition was done in good faith. Accordingly, the devotees' appeal was dismissed.

In *Teng Fuh Holdings Pte. Ltd. v Collector of Land Revenue*,[11] the appellant argued that the decision to acquire was *ultra vires* and in bad faith as first, the land had not been redeveloped for 22 years, and second, the land, which was zoned 'Industrial' when acquired, was later re-zoned to 'Residential'. In light of the conclusive evidence provision, the issue arose as to whether the acquisition could be questioned in Court. The Court of Appeal agreed with the earlier High Court judgment of Phang J. that this was 'not an implausible proposition', and referred to Phang J.'s views on the doctrine of bad faith:

> *Bad faith*, particularly in the *governmental* context, does not sit easily ... [Viewing] the matter from the particular perspective of land acquisition in the Singapore context, it is imperative that a *balance* be found in the tension between ensuring that the purposes of the Act and the ensuing public benefit are achieved on the one hand and ensuring that there is no abuse of power on the other. In this regard, ... the Act was promulgated not only for the public benefit but also because land is an extremely scarce and therefore valuable resource in the Singapore context. These are in fact inextricably related reasons. This being the case, it is clear why much more latitude and flexibility is given to governmental authorities ... but latitude and flexibility stop where abuse of power begins. Such abuse of power is most commonly equated with the concept of *bad faith*. At this point, the courts must – and will – step in.

Phang J. added that 'such abuse of power was a serious allegation, and would not be assumed (let alone be found) at the slightest drop of a hat'. Ultimately, the Court found against the appellant on a technicality – that the application for the order of *certiorari* had been made out of time. Other cases[12] reveal that the *ultra vires* plea has not been successfully pleaded in the Courts. However, this is hardly surprising.

In practical terms, in the absence of public access to most likely confidential details pertaining to Ministerial decisions on acquisition, it would be extremely difficult for a private citizen to raise a 'reasonable suspicion of bad faith'.[13] Hence the unlikelihood of a Court ruling an acquisition decision as *ultra vires*. The next section presents the evidence of public benefit arising from the acquisition powers. In particular, did Mr Lee's vision of a home-owning society materialise?

Evidence of public benefit through densification

The benefits arising from acquisition are evident in diverse sectors in Singapore including industrial and community developments. However, the focus here is on public housing and transport infrastructure. Both are directly related to Singapore's land scarcity issue.

Densification includes the strategy of building a city inwardly by increasing the number of dwelling units and mixed-use spaces per acre. The strategy curbs the tendency for cities to spread out and claim valuable land (Boverket, 2017). The Government's efforts to increase densification in Singapore to create a 'compact city' are most evident in high-rise public housing. With world-class infrastructure for public transport (Urban Redevelopment Authority, 2012a), travel routes and times are shorter, and promote sustainable development by lower car usage and easier access to different activities and locations.

Housing

The most important immediate effect of the legislation was that the government was empowered to acquire large tracts of private land quickly, for the development of industrial facilities for job creation, and new towns for housing. In 1960s, 1.3 million people out of an estimated 1.89 million in Singapore were squatters or residents of slums. Early acquisitions after the formation of the Housing & Development Board (HDB) in 1960s were therefore primarily for public housing (Centre for Liveable Cities, 2014a).

It is evident from Figure 10.2 that the PAP honoured its commitment in 1965 to ramp up housing supply. Indeed, in the 1960s, a flat was being built at the astounding rate of every 45 minutes (Ng, 2008).

By 1985, Singapore had virtually no homeless people, squatters or ghettos (Urban Redevelopment Authority, 2012a). Today, around 83 per cent of Singapore's population of 5.61 million reside in a HDB flat, of which around 90 per cent are homeowners (Housing & Development Board, 2017). The exponential population growth, the number of public housing units and the corresponding drop in building density are depicted in Table 10.1.

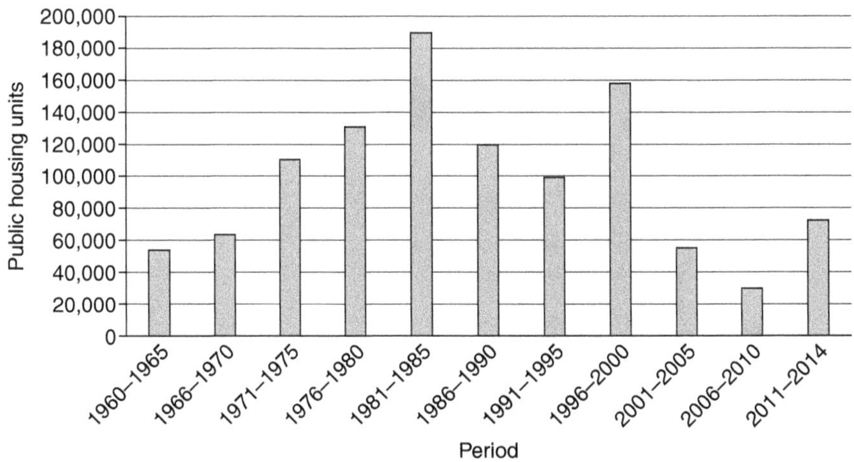

Figure 10.2 Growth in number of public housing units in Singapore.

Source: Housing & Development Board (2015).

Table 10.1 Population and building density in Singapore

Year	*Resident population in public housing under Singapore Improvement Trust* (SIT) and HDB dwellings*	*Number of SIT* and HDB dwellings*	*Building density (persons per dwelling)*
1947	*700,000	*38,500	18.2
1980	1,560,348	320,400	4.87
1990	2,390,200	562,400	4.25
2000	2,978,500	805,000	3.66
2010	3,302,950	943,700	3.45
2016	3,388,525	1,011,500	3.35

Source: author, adapted from Ministry of Trade and Industry Republic of Singapore (2017) and Yuen (2007).

MRT and roads

Acquisition powers were exercised to develop Singapore's infrastructure at an unprecedented rate. These include the construction of the Mass Rapid Transport system (MRT) and roads in Singapore. As can be seen from Figure 10.3, since the first five stations were opened in 1987, the MRT has expanded rapidly to presently 142 MRT stations.

Acquisition for the construction of the North-South Expressway is expected to benefit residents living in the north and north-eastern sectors of Singapore. It will connect at least six major towns to the city-centre (see Figure 10.4). Motorists are expected to be able to save up to 30 per cent of their travel time during peak hours (Land Transport Authority, 2011).

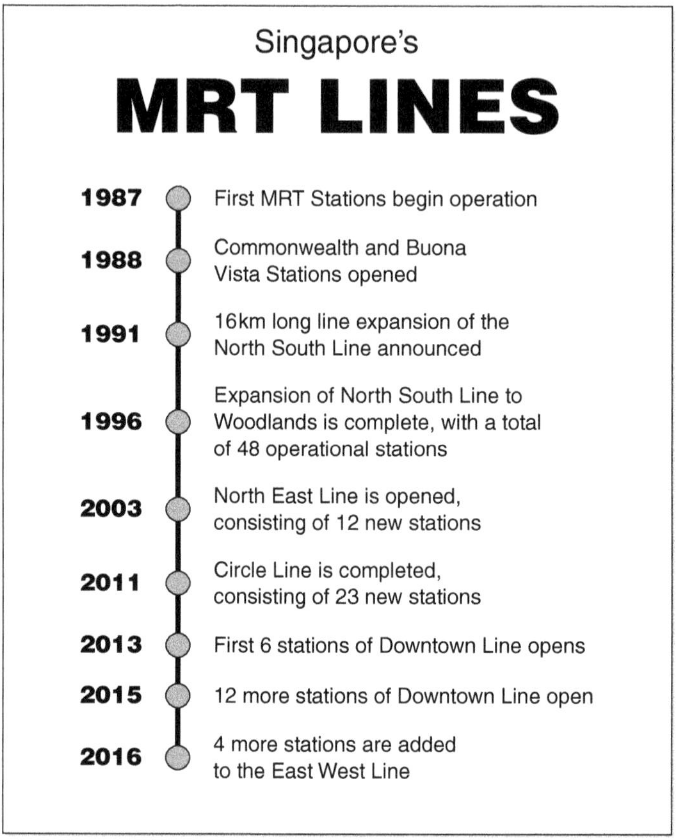

Figure 10.3 Development of MRT network in Singapore.

Source: author's adaptation from Remember Singapore (2018).

Such redevelopment has produced a world-class public transport infrastructure (Sim, 2014). Six out of ten Singapore residents use public transport, even as new train lines expand the country's rail network (Lee, 2016). The increase in the use of public transport is evident from Figure 10.5. Bus and rail ridership rose by 4.3 per cent in 2016 to hit a daily average of 7.2 million – a new record and the 12th consecutive annual rise since 2005 (Tan, 2017a).

Basic tenet underlying compensation for acquisition

Having dealt with the acquisition provisions in the previous section, this section examines the provisions on *compensation* for acquisition. They reflect the underlying principles which determine the 'price' paid by the State and the landowner, to achieve public benefit.

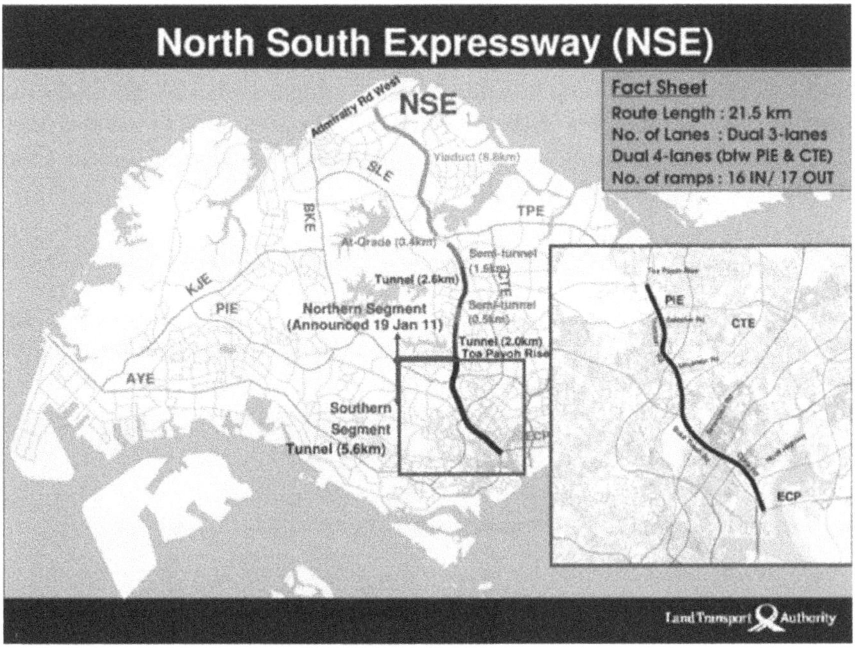

Figure 10.4 The North-South Expressway.

Source: Land Transport Authority (2011).

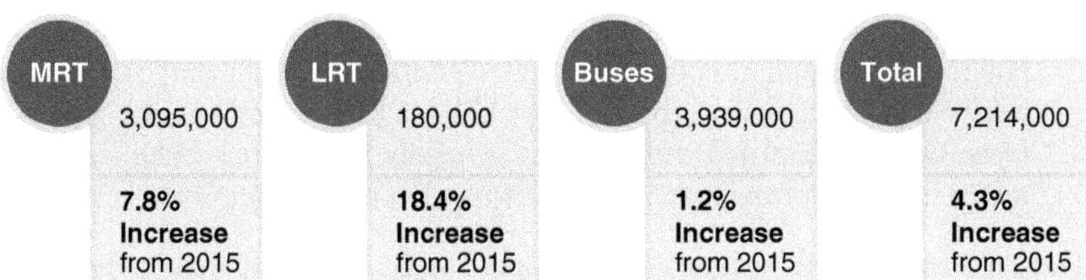

Figure 10.5 Increase in ridership on public transport.

Source: author's adaptation from Tan (2017a).

The forcible dispossession of landowners upon acquisition entitles them to compensation. *Rickets v Metropolitan Rail Co*[14] set out the fundamental 'no worse, no better' principle – that is, compensation for acquisition should put dispossessed landowners in the *same* position as if their property had not been acquired. However, Singapore's compensation provisions reveal clear departures from this principle.

The Collector's inquiry

Upon publication of the President's declaration, the acquisition machinery is triggered. The process culminates with the Collector making the compensation award and taking possession. In *Ezra v Secretary of State for India*,[15] the Collector was described as the agent of the Government. He is required to use the most appropriate valuation method, including the statutory provisions to determine the amount of compensation payable.

'Persons interested'

Only a 'person interested' is entitled to compensation for acquisition. By Section 2(1), this includes 'every person claiming an interest in compensation to be made on account of the acquisition … but does not include a tenant by the month or at will'. Persons interested include companies, owners, mortgagees, reversioners and option holders.

Compensation policy: redistributive justice

The only one basic criterion for judging the worth of public policy proposals is said to be redistribution. The proposed Governmental action should favour the value of doing some good for a community, as opposed to doing great good for an individual (Davidoff, 1975). Mr Lee relied on this principle of redistributive justice as the plank of the compensation framework. He had said (Centre for Liveable Cities, 2014a):

> No private owner should benefit from development that had taken place at public expense, and [the] price paid on acquisition for public purposes should not be higher than what the land would have been worth had the government not contemplated development generally in the area. Such developments should not benefit private landowners but the community at large.

Thus, increases in land values arising from public development were designed to benefit the community and not the landowner. By suppressing compensation, private land for public benefit could be secured without excessive financial burden to the State (and taxpayers).

Compensation provisions: market value and the retrospective statutory date

Section 33(1) requires the Collector to consider the acquired land's 'market value'. The term is not defined, but the Act makes reference to *two* dates for its assessment. These are, the statutory date which is pegged at a retrospective date ('statutory date'), and, the date of the publication of the declaration in the *Gazette*. The date which produces the *lower* market value would be the material date for determining compensation. Section 33(5)(e) required that the potential value of the land be ignored, thereby underscoring the wider public interest of keeping acquisition costs low. In 1973, major amendments capped compensation as at 30 November 1973 levels, or at the date of *Gazette* notification, whichever was lower.

Compensation was to be independent of market conditions or the landowner's purchase price. This statutory date remained for about 14 years. Between 1986 and 1995, the statutory date was amended three times[16] to reflect the increase in property prices over time.

Other provisions affecting compensation

As part of a multi-pronged approach, other provisions were crafted to suppress compensation. This unwavering adherence to the basic tenet of keeping acquisition costs low was explained by Mr Lee as follows (Lee, 2000, 116–117): 'I saw no reason why private landowners should profit from an increase in land value brought about by the economic development and the infrastructure paid for with public funds.' The provisions included the set-off provision, the two-year rule, the seven-year rule, the fire-site and burial grounds provisions. The first three are discussed below.

Set-off provision. Section 33(1)(b) provides the 'set-off' provision applicable in part-lot acquisitions; the Collector was to take into account any increase in the value of any *other* land of the person interested likely to accrue from the use to which the land acquired would be put. To ameliorate the State's burden to pay compensation, the plan was to recoup betterment value arising from Government action. Betterment levy is a form of land value capture (Walters, 2012) which reduces the capital outlay of large public infrastructure developments by requiring landowners to pay for unearned increments. The extent of the set-off could result in reducing the compensation to even nil. For example, in the case of *Chuan Park Condominium* (Chew *et al.*, 2010) the Government acquired the land area of 17 parking lots from the condominium for the construction of an MRT station. An anticipated S$18 million increase in the condominium value arising from the development's future proximity to the station resulted in a nominal compensation of just S$1.

Two-year rule. The 'two-year rule' (section 33(5)(a)) required the Collector to disregard in general, any improvement carried out within two years of any declaration of the intended acquisition. The purpose was to avoid speculation or 'sharp practice' by landowners who could otherwise increase the burden of the State by attempting to enhance their property values.

Seven-year rule. By the 'seven-year rule' (section 33(5)(c)), the benefit of any increment in value arising from the Government's provision of infrastructure such as roads, drains and electricity in the neighbourhood within seven years of acquisition would be discounted when assessing compensation.

Several other provisions reveal the same objective of keeping compensation costs low: for example, by section 34, *inter alia*, the degree of urgency and disinclination of landowner to part with his land were to be *disregarded* when assessing compensation. In light of the labyrinth of constraints which operated to suppress compensation, and the deliberate disregard of critical attributes of the land acquired (such as location and potential use) it was euphemistic to assert that compensation is based on the market value of the land (Khublall, 1994, 129). However, despite all these, there were also several compensation provisions which operated *in favour* of the landowner; they include severance damage, injurious affection, *ex-gratia* payments, reasonable expenses, relocation costs, stamp duty and legal fees (Singapore Land Authority, 2013). The first three are considered below.

Compensation favourable to landowner

Severance damage. This applies to part-lot acquisitions. When only a portion of a person's land is acquired, this may result in diminution in value of the land retained. Section 33(1)(c) provides this shall be taken into consideration in determining the amount of compensation awarded.

Injurious affection. This concept is closely linked to the concept of severance. If the proposed works will injuriously affect the value of the retained land, a claim may be made under section 33(1)(d). As in the case of severance, this provision applies when land is retained by the landowner.

Ex-gratia *payments*. The pegging of compensation at the statutory date gave rise to financial hardship. This was acknowledged by then Law Minister Professor S. Jayakumar, when he said (Ministry of Law, 2007):

> [The] 'whichever is lower' provisions in the present [Act] to cap the statutory compensation have been a source of contention. Over the years, we have sought to cushion the impact of this approach by periodically updating the statutory date and also through *ex-gratia* payments.

Ex-gratia payments were introduced in 1982 to top up the compensation in cases of unusual hardship (Chew *et al.*, 2010). However, this is not an entitlement, nor is there transparency on the criteria used to determine *ex-gratia* payments which are offered on a case by case basis.

Major amendments to compensation provisions in 2007

A watershed date in the legal history of acquisition in Singapore is 12 February 2007. It marked the shift away from the statutory date, to current market value. Professor Jayakumar stated that the statutory date and the 'whichever is lower provisions' were 'no longer appropriate in the current context' (Ministry of Law, 2007). The valuation for compensation would have to take into account the land's permitted use and the potential value realisable under the Master Plan. Compensation would be based on the open market value which a *bona fide* purchaser would reasonably pay for the land. Other major amendments included the abolition of the seven-year and two-year rules.

Appeals against compensation award

The extensive ambit of the Government's acquisition powers, and relatively bleak prospects for objections or challenges to acquisition have been discussed above. However, fortunately, there is a clear path for the landowner to object and appeal against the *compensation* awarded. The compensation award is only an offer, and not binding on the landowner. By section 23(1), the landowner may serve a notice of appeal to the Appeals Board. The Appeals Board is an *administrative* tribunal entrusted with matters of specialist evaluation by persons with requisite expertise (Singapore Parliament, 1966). Yet the proceedings are *judicial* in nature, unlike the Collector's inquiry. A summary of the procedure is outlined in Figure 10.6.

Figure 10.7 illustrates the number of appeal cases which successfully challenged the compensation award. Two matters are evident: first, the number of appeal cases declined over time; second, after 2007, there is a significant decrease in the number of *successful* appeal cases. These coincide with the revision to the statutory dates at closer intervals (Chew *et al.*, 2010), and post-2007 amendments respectively.

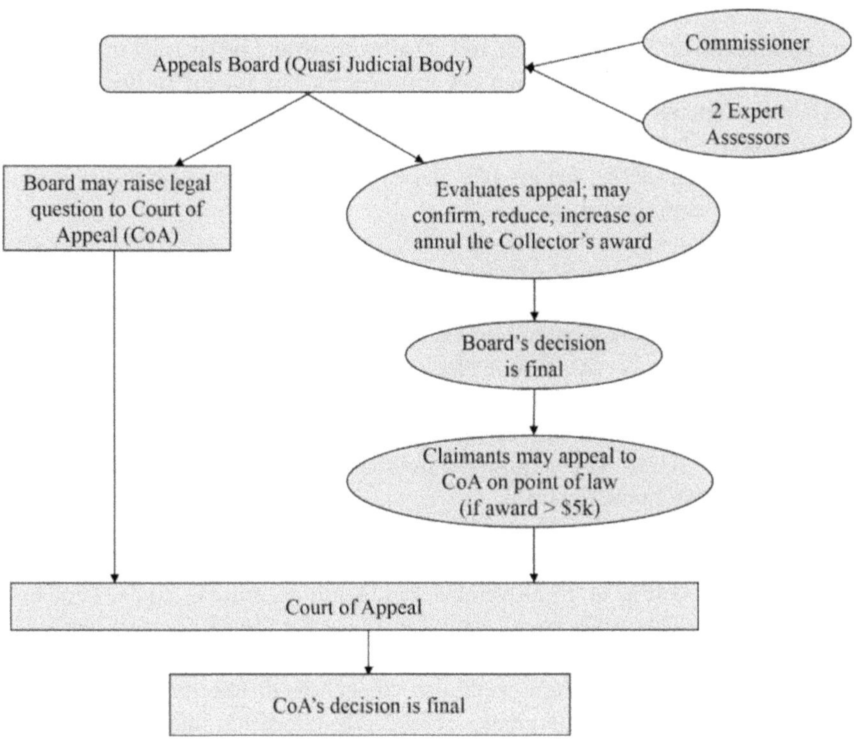

Figure 10.6 Procedure for appeal against compensation.

Source: author's adaptation from Part III of the Act.

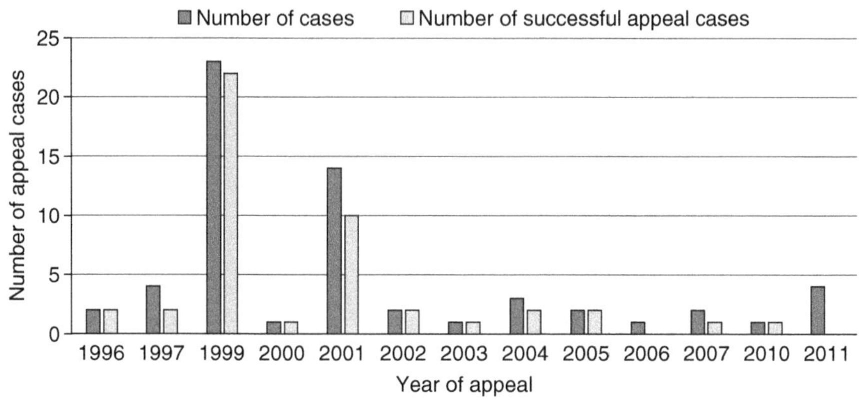

Figure 10.7 Appeals against compensation award.

Source: author's compilation.

Amendments in 2014 and 2015 *inter alia* removed the betterment levy. In part-lot acquisitions, the potential increase in the value of the owner's remaining land would no longer be deducted. The owner would be entitled to compensation at full market value. This move was described as '[being] in line with the 2007 amendments to peg the compensation of acquired land to the market value which a *bona fide* purchaser would reasonably be willing to pay for the property' (Rajah, 2014). Figure 10.8 depicts the evolution of acquisition and other related legislation to date.

Softening the blow: incentives for aggrieved landowners

This section considers some *other* ways in which aggrieved landowners were placated. In November 2011, four blocks of 39-year-old public housing flats in Rochor Centre were slated to make way for the 21.5km-long North-South Expressway which would run beneath some of the busiest parts of the city (Straits Times, 2011). The acquisition involved 567 flats, 187 shops and eating houses and 3 communal facilities.

Affected residents were compensated at prevailing market rates, given the option to purchase new HDB flats built near the Kallang MRT station at subsidised prices (frozen as at 15 November 2011) and accorded a 20 per cent discount. Alternatively, they could apply for a new flat elsewhere with the same relocation benefits, or sell their existing flat in the open market (Straits Times, 2017). Displaced shop tenants were also compensated. Enterprises continuing their businesses were given an extra S$30,000 in relocation help and also entitled to a 10 per cent discount for HDB rental shops. The Rochor acquisition attracted much media attention, particularly because there was no consultation before the acquisition notice, and probably because most displaced owners

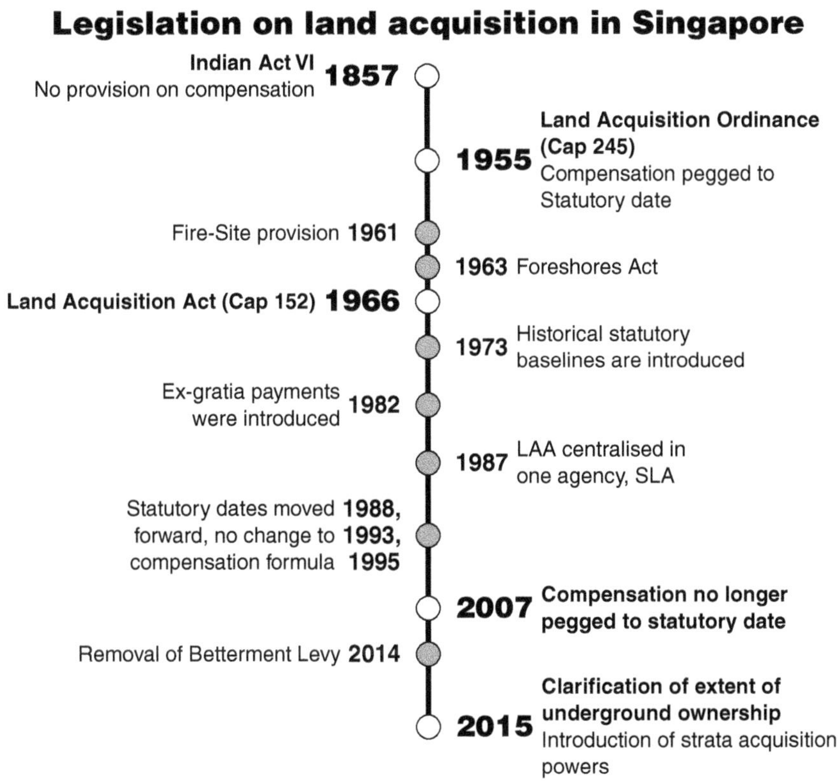

Figure 10.8 Land acquisition legislation in Singapore.

Source: author.

comprised the elderly. If their needs were not satisfactorily addressed, they would suffer disproportionate hardship and this would reflect poorly on the acquisition and resettlement process.

Other related legislation: the Foreshores Act

Between 1960 and 2007, the State's ownership of land doubled from 44 per cent to over 85 per cent (Centre for Liveable Cities, 2014a). One reason was extensive land reclamation, which grew Singapore by 22 per cent of its original area (The Economist, 2015). This was made possible by legislation which facilitated acquisition for land reclamation.

Acquisition for land reclamation

In 1970, for economic and environmental reasons, as well as land maximisation, the Government decided to relocate the main airport from Paya Lebar to

Changi. Changi Airport was to be built on reclaimed land, with a runway butting out to the sea, just like Hong Kong's Kai Tak Airport. The flight path both to the north and south would be largely over water, thereby addressing the noise issue faced at Paya Lebar, where the flight path was directly over densely populated residential areas. This idea was inspired by Mr Lee's visit to Boston's Logan Airport, where planes took off and landed over water, reducing aircraft noise (Urban Redevelopment Authority, 2012a).

Also, Mr Lee believed that to be truly world-class as a hub for business and tourism, the city needed to be easily and quickly accessible from the airport. Thus the Government decided to build an expressway to connect Changi Airport to the Central Business District. However, as the entire east coast was already built up to the coastline, there was no land (Urban Redevelopment Authority, 2013, 5).

In addition, the high cost of compensation for coastal land posed a real obstacle to reclamation. Mr Lee recounted (Urban Redevelopment Authority, 2013, 5) how he dealt with it: '[We] passed a law that said that when government acquires coastal land, we compensate without taking into account that it's by the seaside.' The Foreshores Act was accordingly amended to acquire as much land as possible and reclaim coastal areas, *with no right to compensation on account of lost sea frontage*.[17] The entire coastline was then developed for infrastructure, recreation and *public* housing on prime, reclaimed seafront land which in many countries, would have been exclusively owned by the wealthy (Centre for Liveable Cities, 2014a).

Summary and conclusion

Alongside public sector redevelopment for a variety of purposes through acquisition, the densification effort was further facilitated through *private sector* participation. Radical legislation passed in 1999 enables developers to realise the redevelopment potential of under-utilised strata-titled land parcels (Christudason, 2012). This has resulted in the rejuvenation, densification and optimisation of under-utilised land parcels, yielding more than 35,000 new strata units. In 2017 alone, 14 collective sales were transacted, with the potential of yielding 17,546 units (Propertyinvestsg, 2017). However, the focus of this chapter has been densification through *public* redevelopment led by the Government.

Through the lens of Singapore's legal framework for acquisition, this chapter has examined one of the main reasons for Singapore's dramatic transformation from a third- to first-world country within just 50 years. The theoretical justification for acquisition for public redevelopment is that the welfare of the public is paramount, and that public necessity is greater than private. This is authenticated in Singapore's context. Acquisition was carried out for a broad range of purposes and clearly yielded public benefit: today more than 80 per cent of the population own and reside in public housing with easy access to amenities and public transportation while industrialisation created much-needed jobs and became a key driver of the economy. None of this would have been possible without the legal framework for acquisition.

As succinctly stated by the Law Minister (Ministry of Law, 2007):

> The ... [Act] has played a crucial role in the development of Singapore. It paved the way for numerous major developments in Singapore, including public housing for the vast majority of the population, the Jurong Industrial Estate which provided jobs for many, key infrastructure such as the Changi International Airport and the various MRT and road projects. These ... brought much benefits in the public interest, [and] were made possible because they were developed at a cost which was affordable to the State then, through the various limiting compensation provisions under the Land Acquisition Act.

Criticisms have been levelled at not only the 'draconian' nature of the provisions, but about the implementation process. One example was the acquisition of Bukit Brown Cemetery in 2011. The government faced public outcry regarding plans to build an eight-lane road which would cut into the cemetery. Raw nerves were touched as five thousand graves of forefathers had to be exhumed to make way for the construction (Singapore Heritage Society, 2012). Again, there was no consultation prior to the acquisition notice. During the Parliamentary debates (Asiaone.com, 2012), emotional last-ditch attempts were made to save Bukit Brown. However, then Minister of State for National Development, Mr Tan Chuan-Jin explained that the Government had already explored various other options, before it made the difficult decision to build the proposed road. Mr Tan also recounted the following incident. In the 1960s, PAP's first Minister for National Development Mr Lim Kim San ('Mr. Lim', also known as 'Mr HDB') (Ng, 2008) had faced similar sentimental concerns about exhumations to make way for public housing at Tiong Bahru. When an aggrieved group went to see him, Mr Lim responded with this rhetoric, 'do you want me to look after our dead grandparents, or do you want to look after your grandchildren?'

Another example relates to the construction of the Cross Island MRT Line beneath Singapore's largest nature reserve. Critics say this is actually counterproductive as it will traverse an unpopulated, forested area, and also inflict irreparable damage to a pristine habitat (Tan, 2017b). Yet despite the above and other criticisms (Latif, 2009), the transformation achieved through the acquisition mechanism is nothing short of spectacular. These reasons have been cited for its success (Phang and Helble, 2016):

i Legal and constitutional mandate as well as clear processes in the form of the LAA;
ii Meticulous processes for record keeping and calculation of compensation;
iii The superiority of alternative accommodation, business premises and environments offered by the government to affected persons; and
iv Effective forward planning and coordination to ensure smooth resettlement.

Meanwhile, there are future acquisition plans for Singapore to expand the Tengah Air Base (National Environment Agency, 2017). Land occupied by fish farms and a nursery will be acquired and more than 80,000 graves at Choa Chu Kang Cemetery will be exhumed. Hopefully, the government will approach the matter with greater sensitivity to avoid a repeat of the public's outcry over the Bukit Brown Cemetery acquisition.

Singapore's dramatic physical transformation was neither fortuitous nor kismet. Rather, the founding fathers, led by Mr Lee, had a long-term development vision which they combined with a resolute will to translate into reality. The Government's focused and hard-nosed approach was matched closely by meticulous legal measures and close coordination between agencies to implement acquisition, resettlement and redevelopment. Today there is even a 'landscape replacement policy' to ensure that greenery lost from site areas taken up by buildings would be replaced (Urban Redevelopment Authority, 2012a). Timely and demonstrable improvements in living standards were key to garnering broad-based public support (Urban Redevelopment Authority, 2012b). The single-mindedness with which the HDB took on the public housing agenda is summed up here: '*you cannot make an omelette without breaking eggs*' (Heng and Aljunied, 2009, 93). The nation of sojourners became a nation of stayers (Straits Times, 2017).

Notes

1 Cap. 152 of the Revised Edition of Singapore Statutes 1985. All references are to provisions of this Act unless otherwise stated.
2 This refers to compulsory enlistment for male Singaporeans under the Enlistment Act, Cap. 93, Revised Edition of Singapore Statutes, 2001.
3 All India Reports (1914) PC 20.
4 The rates for this tax are set on a bi-annual basis by Singapore's Chief Valuer through a Table of Rates and a sector map. Accessed 16 September 2017: www.ura.gov.sg/uol/guidelines/development-control/development-charge.
5 [1981] 1 *Malayan Law Journal* 9.
6 *Chiltern v Telford Development Corporation* [1987] 3 All England Reports 992 at p. 997.
7 *R v Secretary of State ex p de Rothschild* [1989] 1 All England Reports 933 at p. 935.
8 Now referred to as a 'Quashing Order': Order 53, Supreme Court of Judicature Act, Cap. 322. Revised Edition of Singapore Statutes 1985.
9 [1989] 2 SLR 52.
10 [2009] 2 SLR 542; [2009] SGCA 1.
11 [2007] SGCA 14.
12 Including *Galstaun v Attorney General op. cit* and *Kentucky Fried Chicken Management Pte. Ltd v The Attorney-General & Anor.* [1992] SGHC 225.
13 *Teng Fuh Holdings Pte. Ltd. v Collector of Land Revenue* [2007] 2 SLR 568; [2007] SGCA 14 at para. 27.
14 [1867] LR HL 175.
15 [1905] ILR Cal 605.
16 The dates are 30 November 1973, 1 January 1986, 1 January 1992 and 1 January 1995. Only one of these dates is relevant, depending on the date of the acquisition.
17 Emphasis mine. See also section 7 of the Foreshores Act (Cap. 113), Revised Edition of Singapore Statutes 1985.

References

Asiaone.com (2012) Tan Chuan-Jin vs Janice Koh on Bukit Brown, 6 March. Accessed 9 September 2017 at www.asiaone.com/print/News/Latest%2BNews/Singapore/Story/A1Story20120306-331785.html.

Boverket, Swedish National Board of Planning, Building and Housing (2017) *Urban Density Done Right: Ideas on Densification of Cities and Other Communities*. Accessed 19 September 2017 at www.boverket.se/globalassets/publikationer/dokument/2017/urban-density-done-right.pdf.

Centre for Liveable Cities (2014a) *Land Acquisition and Resettlement: Securing Resources for Development. Urban Systems Studies*. Accessed 11 September 2017 at https://smartnet.niua.org/sites/default/files/resources/USS_Land_Acquisition_and_Resettlement.pdf.

Centre for Liveable Cities (2014b) *Financing a City: Developing Foundations for Sustainable Growth*. Accessed 15 September 2017 at www.clc.gov.sg/documents/uss/USS_Financing_a_city.pdf.

Centre for Liveable Cities (2015) *Built by Singapore: From Slums to a Sustainable Built Environment*. Accessed 10 September 2017 at www.clc.gov.sg/documents/uss/USS_Built_by_Singapore.pdf.

Changi Airport Group (2017) Traffic statistics. Accessed 17 September, 2017 at www.changiairport.com/corporate/about-us/traffic-statistics.html.

Chew B, Hoong V, Tay L K and Vellasamy M (2010) Compulsory acquisition of land in Singapore: A fair regime? *Singapore Academy of Law Journal*, 22, 166–188.

Christudason A (2012) Urban rejuvenation through collective (*enbloc*) sales in Singapore: Property rights or property wrongs? *Journal of Urban Regeneration and Renewal*, 5(1), 51–64. ScholarBank@NUS Repository. Accessed 9 September 2017 at http://scholarbank.nus.edu.sg/handle/10635/46280.

Davidoff P (1975) Working toward redistributive justice, *Journal of the American Institute of Planners*, 41(5), 317–318.

Gurran N and Lawler S (2016) Explainer: What is 'value capture' and what does it mean for cities? Accessed 15 September 2017 at http://theconversation.com/explainer-what-is-value-capture-and-what-does-it-mean-for-cities-58776.

Heng D and Aljunied S M K (2009) *Reframing Singapore: Memory – Identity – Trans Regionalism*. Amsterdam: ICAS Publications Series. Amsterdam University Press.

Housing & Development Board (2015) *HDB Annual Report 2014/2015*. Singapore: Housing Development Board.

Housing & Development Board (2017) *Public Housing – A Singapore Icon*. Accessed 10 September 2017 at www.hdb.gov.sg/cs/infoweb/about-us/our-role/public-housing-a-singapore-icon.

Hui E C M, Ho V S M and Ho D K H (2004) Land value capture mechanisms in Hong Kong and Singapore: A comparative analysis, *Journal of Property Investment and Finance*, 22(1), 76–100.

Khublall N (1994) *Compulsory Land Acquisition in Singapore and Malaysia*, 2nd ed. Singapore: Butterworths Asia.

Land Transport Authority (2011) Full alignment of north-south expressway unveiled. Accessed 10 September 2017 at www.lta.gov.sg/apps/news/page.aspx?c=2&id=mjtut44y69ma2c6354z6ob21shwo5gewr7tj44yhiqvkx240r0.

Latif A (2009) *Lim Kim San: A Builder of Singapore*. Singapore: Institute of Southeast Asian Studies, Singapore.

Lee K Y (2000) *From Third World to First: The Singapore Story 1965–2000*. Singapore: Singapore Press Holdings.

Lee P (2016) More Singaporeans take bus, MRT to work: Government survey, *Straits Times*, 9 March. Accessed 7 September 2017 at www.straitstimes.com/singapore/more-singaporeans-take-bus-mrt-to-work-government-survey.

Lim C J (1968) Compulsory land acquisition in Singapore, *Malaya Law Review*, 10(1), 28.

Loh K S (2008) Fires and the social politics of nation-building in Singapore, Asia Research Centre Working Paper no. 149. Accessed 7 September 2017 at www.murdoch.edu.au/Research-capabilities/Asia-Research-Centre/_document/working-papers/wp149.pdf.

Loo E (2017) Lessons from Singapore about land value capture. Accessed 17 September 2017 at www.rtpi.org.uk/briefing-room/rtpi-blog/lessons-from-singapore-about-land-value-capture/.

Mahbubani K (2015) Why Singapore is the world's most successful society. Accessed 10 September 2017 at www.mahbubani.net/articles%20by%20dean/Why%20Singapore%20Is%20the%20World's%20Most%20Successful%20Society%C2%A0_%C2%A0Kishore%20Mahbubani.pdf.

Ministry of Culture (1966) National Archives of Singapore. Speech by the Minister of Law, E W Barker moving the Second Reading of the Land Acquisition Bill in the Singapore parliament. Accessed 1 October 2017 at www.nas.gov.sg/archivesonline/data/pdfdoc/PressR19660622e.pdf.

Ministry of Law (2007) Second Reading Speech on Land Acquisition (Amendment) Bill by Deputy Prime Minister Professor S. Jayakumar. Accessed 9 September 2017 at www.mlaw.gov.sg/news/parliamentary-speeches-and-responses/second-reading-speech-on-land-acquisition-amendment-bill-by-dpm-prof-s-jayakumar-11-apr-2007.html.

Ministry of Trade and Industry Republic of Singapore (2017) *Population Trends 2017*. Singapore: Department of Statistics.

National Environmental Agency (2017) *Tengah Air Base to Be Expanded to Free up Paya Lebar Air Base for Future Development*. Singapore: National Environmental Agency.

Ng Y L (2008) Mr Lim Kim San (1916–2006) – 'Mr HDB'. Accessed 18 September 2017 at www.nas.gov.sg/archivesonline/html/etc/lim_kim_san.htm.

Ong C (2016) Albert Winsemius. Accessed 10 September 2017 at http://eresources.nlb.gov.sg/infopedia/articles/SIP_1457_2009-02-11.html.

Phang S Y and Helble M (2016) Housing policies in Singapore. In: *The housing challenge in emerging Asia*, 174–209. Research Collection School of Economics. Accessed 12 September 2017 at http://ink.library.smu.edu.sg/soe_research_all/20.

Planning, Environment and Lands Branch (1996) Urban renewal in Hong Kong. Accessed 14 September 2017 at www.cityu.edu.hk/hkhousing/pdoc/1996_Urban-RenewalinHongKong.pdf.

Propertyinvestsg (2017) Enbloc market updates. Accessed 8 September 2017 at www.propertyinvestsg.com/en-bloc-market-updates-2017-sept/.

Rajah I (2014) Enhancing land acquisition compensation framework: Land Acquisition (Amendment) Bill 2014. Accessed 30 August 2017 at www.mlaw.gov.sg/content/dam/minlaw/corp/News/LAA%202014%20Newsletter.pdf.

Remember Singapore (2018) A journey into Singapore's MRT past. Accessed 12 September 2017 at https://remembersingapore.org/2013/7/23/singapore-mrt-history/.

Sim R (2012) Pearls Centre tenants surprised by acquisition for new Thomson line, *Straits Times*, 30 August. Accessed 9 September 2017 at www.straitstimes.com/singapore/transport/pearls-centre-tenants-surprised-by-acquisition-for-new-thomson-line.

Sim R (2014) Singapore's public transport system one of world's most efficient, *Straits Times*, 2 July. Accessed 10 September 2017 at www.straitstimes.com/singapore/transport/study-singapores-public-transport-system-one-of-worlds-most-efficient.

Singapore Heritage Society (2012) Position paper on Bukit Brown. Accessed 10 September 2017 at www.singaporeheritage.org/wp-content/uploads/2011/11/SHS_BB_Position_Paper.pdf.

Singapore Land Authority (2013) Pearls Centre owners issued compensation and given additional assistance to relocate. Accessed 27 August 2017 at www.sla.gov.sg/News/articleid/56/parentId/97/year/2013?category=Press%20Releases.

Singapore Parliament (1966) *Report of the Select Committee on the Land Acquisition Bill* at C31. Singapore: Tan Boon Poh, Government Printer.

Straits Times (2011) More than 500 homes to make way for highway, 15 November. Accessed 10 September 2017 at www.straitstimes.com/singapore/transport/more-than-500-homes-to-make-way-for-highway.

Straits Times (2017) What makes Singapore brand tick?, 19 September. Accessed 19 September 2017 at www.straitstimes.com/opinion/st-editorial/what-makes-singapore-brand-tick.

Tan C (2017a) Bus, rail ridership soars to new high, *Straits Times*, 22 April. Accessed 9 September 2017 at www.straitstimes.com/singapore/transport/bus-rail-ridership-soars-to-new-high.

Tan C (2017b) Cross Island Line debate misses elephant in room, *Straits Times*, 16 February. Accessed 9 September 2017 at www.straitstimes.com/opinion/cross-island-line-debate-misses-elephant-in-room.

The Economist (2015) Such quantities of sand, 26 February. Accessed 12 September 2017 at www.economist.com/news/asia/21645221-asias-mania-reclaiming-land-sea-spawns-mounting-problems-such-quantities-sand.

Urban Redevelopment Authority (2012a) Designing our city: Planning for a sustainable Singapore. Accessed 14 September 2017 at www.ura.gov.sg/skyline/skyline12/skyline12-03/special/URA_Designing%20our%20City%20Supplement_July12.pdf.

Urban Redevelopment Authority (2012b) Foreword by Vincent Hoong, Chief Executive Officer, Singapore Land Authority. Urban systems studies. Land acquisition and resettlement: Securing resources for resettlement. Accessed 5 September 2017 at https://smartnet.niua.org/sites/default/files/resources/USS_Land_Acquisition_and_Resettlement.pdf.

Urban Redevelopment Authority (2013) Remarks by Mr Peter Ho, chairman of the Urban Redevelopment Authority at the 'Lee Kuan Yew and the physical transformation of Singapore' public conference. Accessed 18 September 2017 at www.clc.gov.sg/documents/otherevents/leekuanyewandthephysicaltransformationofsingapore_high lights.pdf.

Walters C L (2012) Land value capture in policy and practice. Accessed 15 September 2017 at www.landandpoverty.com/agenda/pdfs/paper/walters_full_paper.pdf.

Yuen B (2007) Squatters no more: Singapore social housing. *Global Urban Development* 3(1). Accessed 9 September 2017 at www.globalurban.org/GUDMag07Vol. 3Iss1/Yuen%20PDF.pdf.

11 Everybody needs good neighbours, especially in strata

Are new Australian laws enabling forced sales of strata properties justified?

Laura Crommelin, Laurence Troy, Bill Randolph and Hazel Easthope

Introduction

Since the introduction of strata title in 1961 – which allows "the subdivision of land and airspace into individually owned freehold apartments and collectively owned common property" (Sherry, 2017, 9) – the Australian state of New South Wales (NSW) has seen a dramatic increase in high density living. More than a quarter of Sydneysiders now live in strata titled properties, and this number continues to grow. Over the past three decades, this urban transformation has been strongly promoted by metropolitan planners and policy makers, who have sought to encourage consolidation and densification of existing urban areas. This 'compact city' model of urban planning has also received significant support from the development industry, which has worked hard to shift public attitudes towards high density living.

The rollout of this urban transformation strategy, however, brings new challenges. As the supply of underutilised land in existing residential areas dries up, urban renewal efforts must shift to the more complex task of replacing existing housing. Many early strata schemes – particularly in inner and middle ring suburbs and near transport – now occupy land under increasing pressure from planners, politicians and developers for more intensive use. Furthermore, many are ageing and require substantial reinvestment to upgrade their facilities and address safety concerns.

Faced with this dilemma, jurisdictions worldwide are changing their laws to facilitate the redevelopment of multi-unit blocks. The primary strategy involves removing the requirement for all owners to approve the termination of a strata scheme, which is a prerequisite to redeveloping the building. By removing the need for unanimous consent, these laws mean that if a large majority of owners in a scheme approve, the minority of dissenting owners can be forced to sell their properties.

NSW is one of the first Australian jurisdictions to introduce this change, with new laws in effect since July 2016. This chapter outlines the nature of these changes, and considers whether they are legally and socially justifiable.

It does so by highlighting how the new strata termination model raises many of the same legal and social issues as compulsory acquisition laws, but offers fewer protections to dissenting owners. In addition, it outlines some public reactions to the issues raised by the new termination model, drawing on the results of a multi-year, mixed-methods research study conducted in the lead-up to the legislative change. A collaboration between the authors and key stakeholders from industry and government, the study involved 34 interviews with strata experts, an economic analysis of the feasibility of strata renewal across Sydney, a survey of 1261 strata owners and residents, and 6 community focus groups (see Troy *et al.*, 2015a for details).

The findings demonstrate that many strata residents have a nuanced understanding of the need to balance multiple sets of rights in strata. While not opposed to changes to prevent unfair profit seeking by 'hold out' owners, most participants were keen to ensure that vulnerable people were protected, and that renewal serves a public benefit. In the light of these findings, the chapter concludes by highlighting the risks if strata owners are not granted protections similar to those offered by the compulsory acquisition regime, and by arguing that strata termination laws need to be made more equitable.

The new strata termination legislation

In October 2015 the New South Wales Parliament passed two new Acts replacing all existing strata legislation, which came into force in July 2016. Of particular interest for this chapter is Part 10 of the *Strata Schemes Development Act 2015* (the Act), which reduces the level of owner approval required to terminate a scheme to 75 per cent.[1] Termination can occur in two circumstances – either by collective sale to a third party (e.g. a developer), or by collective renewal, where the scheme is terminated and existing lot owners become tenants in common (shared owners) of the property. The latter enables existing lot owners to retain their property rights and develop a new scheme collectively.

The process set out in the Act for achieving a termination in either circumstance is as follows:

- A proposal for the collective sale or renewal of the building is given to the strata committee (SC).
- The SC chooses whether to put the proposal to the owners' corporation (OC) to decide whether to investigate the proposal further.
- If the proposal is put to the OC and it votes to proceed, a strata renewal committee is formed (comprised of OC members) and is given up to a year to develop a full strata renewal plan, outlining how the collective sale or renewal would take place.
- Once the plan is completed, owners vote on whether to approve the plan. Each lot (unit) owner has one vote, and 75 per cent of lot owners must vote in favour for the process to proceed.

- If approved, the OC must also pass a special resolution (requiring 75 per cent of unit entitlement votes) to send the plan to the NSW Land and Environment Court for approval.
- The Court must review the plan and determine that "the terms of the settlement are just and equitable in all the circumstances" before making an order approving the plan.
- If the Court makes such an order, any dissenting owners must sell their lots, and will be paid the greater of the agreed price (as divided between all lot owners based on unit entitlements) or the compensation amount, determined by applying the 'just terms' test in existing compulsory acquisition legislation.

While NSW is not the first jurisdiction to remove the unanimous consent requirement, the 75 per cent threshold is relatively low (Table 11.1 shows the thresholds in some other key jurisdictions). On the other hand, the NSW approach requires two votes to proceed – 75 per cent of all owners, and 75 per cent of votes based on unit entitlements – whereas others only require one vote to approve a termination.

Other Australian jurisdictions are now following NSW's lead, with Queensland and Western Australia both proposing to introduce similar thresholds (Commercial and Property Law Research Centre and QUT Law, 2014; Landgate, 2014). For this reason, it is important that the issues associated with these legislative reforms are widely recognised, as more Australians may soon find themselves facing the same challenges as NSW's strata residents.

How similar is strata termination to compulsory acquisition?

One way to assess the new strata termination process is to consider how it aligns with compulsory acquisition by state agencies, and whether it provides

Table 11.1 Strata termination thresholds in key jurisdictions

Jurisdiction	*Threshold*	*Relevant legislation*
Singapore	80%[2]	*Land Titles (Strata) Act* Cap. 158 S.84A
Hong Kong	90%[3]	*Land (Compulsory Sale for Redevelopment) Ordinance* (Cap. 545) 1998
British Columbia	80%	*Strata Property Act* 1998, s.272
Florida	80%	*Florida Condominium Act* §718.117
New Zealand	75%	*Unit Titles Act 2010*, §187–189
Northern Territory	By development age: 100% for <15 years 15–<20 years = 95% 20–<30 years = 90% 80% for 30 years plus	*Termination of Units Plans and Unit Title Schemes Act* 2014

dissenting owners with similar protections against unjust treatment (Sherry, 2015). In Australia, the Federal Government is empowered to compulsorily acquire land under s.51(xxxi) of the Constitution, which permits "the acquisition of property on just terms from any State or person for any purpose in respect of which the Parliament has powers to make laws". The situation at the state level is different, however, as Russell (2014, 14) explains:

> the State Parliaments have no constitutional restriction in relation to acquisition of land and have no constitutional obligation to provide any compensation whatsoever for such compulsory acquisition ... Each State Parliament may enact legislation to compulsorily acquire land with or without the payment of compensation.

All states and territories have enacted compulsory acquisition laws, and while these vary by jurisdiction, all provide for compensation to be paid (Russell, 2014). In NSW, the relevant legislation is the *Land Acquisition (Just Terms Compensation) Act 1991*, which provides for compensation at not less than market value.

Importantly, while there is also no constitutional requirement to acquire land only for public purposes, Sherry (2015) notes that "in order to avoid revolution, state governments ... restrict resumptions to projects that have some (even if arguable) public benefit". Government documents confirm this, stating that "[i]n NSW all levels of government can acquire privately owned land for public purposes" (Valuer-General's website, www.valuergeneral.nsw.gov.au/compulsory_acquisitions). This includes local councils, whose powers of compulsory acquisition are subject to additional restrictions under the *Local Government Act 1993* (NSW), particularly where the land acquired is to be resold.

This cautious approach is reflected in Australian jurisprudence on compulsory acquisition, with courts generally interpreting these legislative powers narrowly. In particular, Australian courts have been reluctant to allow the exercise of compulsory acquisition powers if a private entity will benefit financially from the outcome, even if public benefit may also result (Pocock, 2014; 2015). This includes the use of compulsory acquisition to achieve urban renewal outcomes like the redevelopment of Parramatta's civic square (see Chapter 12). In that case, the Court prevented the acquisition of land because it would subsequently be transferred to a private developer involved in a public-private partnership with the local municipal government (see Pocock, 2014 for a detailed review of the legal reasoning).

It is worth noting that Australia is an outlier in refusing to allow such acquisitions. In the US (where compulsory acquisition is known as 'eminent domain') they are referred to as 'economic development takings' (see Chapter 3), while in the UK such acquisition is enabled under the *Town and Country Planning Act 1990* (UK) Pt IX (see Chapter 4). While some scholars argue that compact city imperatives now justify the adoption of a more flexible approach in Australia, others note the history of eminent domain highlights the problems this can cause (see Pocock, 2014, 2015 for both sides of the debate). Of course, questions also remain about the legitimacy and effectiveness of such compact city imperatives more broadly (see Searle, 2004; Bunker *et al.*, 2017), which become even more pressing when involuntary acquisition of property is required to achieve these imperatives.

Furthermore, the US experience shows how poor and vulnerable groups are particularly disadvantaged in this process, while more politically connected groups have used their influence to achieve beneficial outcomes. Examples include corporate entities like General Motors or Pfizer lobbying for the use of eminent domain to achieve urban renewal outcomes that benefit them directly (Garnett, 2006). Poor neighbourhoods with high rates of lower-income and minority groups have borne the brunt of the dislocation caused by these problematic urban renewal projects.

Despite the different legal and political context, these outcomes are of interest in relation to Australian strata renewal, as our research indicates vulnerable groups, including lower-income, elderly and culturally and linguistically diverse people, will be negatively affected by the new termination process. Our research (Troy *et al.*, 2015b) demonstrates that, in Sydney at least, leaving strata redevelopment to the private market will result in uneven outcomes, and will often contribute to poor housing outcomes for vulnerable people. A detailed feasibility analysis of strata redevelopment identified three main scenarios:

i **Gentrification**: In high-value areas near the city centre and coast, the likely outcome is redevelopment that improves building quality without necessarily increasing density – i.e. development of new and more expensive buildings of a similar size. This scenario is likely to displace moderate and lower-income residents, especially renters and elderly owners, whose older apartments would be unlikely to fetch a high enough price to allow them to buy back into the highly pressured local market.

ii **Densification**: In less expensive middle-ring suburbs, the likely outcome is increased density but without necessarily increasing building quality. While this serves the public purpose of increasing housing supply, there was significant concern among our research participants about the likely quality of the new apartment development in these scenarios, as well as the additional pressure on local services and infrastructure.

iii **Residualisation**: In the outer western suburbs, where many older and poorly maintained apartment buildings exist, the much-needed upgrading is unlikely to occur through private market mechanisms alone. The feasibility analysis showed there is insufficient value in the land to generate a market response to the growing problem of low-income households living in poorly maintained and deteriorating apartment blocks in these areas.

Overall, the feasibility analysis suggests that while market-driven renewal of older multi-unit apartment buildings will occur in parts of Sydney, it is unlikely to occur in a way that provides a public benefit by meaningfully addressing housing supply shortages or poor housing conditions. Compounding these concerns is the fact that there is significant variation in the levels of political and social capital of strata residents across Australian cities. This was particularly evident from the questions and concerns raised at our community forums in Cabramatta (a lower-income suburb with a large migrant population) and Mosman (one of Sydney's wealthiest suburbs), which highlighted the difference in knowledge and access to expert advice about strata title ownership that these communities enjoy. In many cases, this means that lower-income and vulnerable groups are less likely to share in any benefits arising from urban renewal through strata scheme redevelopment.

Furthermore, the new Act does not include any requirement that a non-unanimous termination serve some public purpose – be it either the improvement of a dangerous or decrepit building, or an increase in residential density. This is despite the former purpose being a stated justification for the legislative change, with the responsible minister claiming it was "designed to empower strata owners to make a collective decision about the most important issue that will confront all strata buildings at some point: what to do with the building as it ages" (NSW Parliament, 2015). While the government largely focused on the issue of building disrepair in promoting the new strata renewal process, one state government interviewee did acknowledge that broader planning objectives were also a driver: "Obviously, some areas which are heavily strata-fied [*sic*] require either significant upgrading or refurbishment but also, as is eloquently positioned in part of the scoping papers here,[4] there requires some densification around these high value areas."

Comments from the property industry suggest they also saw broad renewal goals as a driver for the change, with one industry representative describing the new process as "breaking the 100 per cent deadlock that has prevented tired old buildings from being refreshed and *stifled vital urban renewal*" (Property Council, 2015, our emphasis).

Nonetheless, these public purposes are notably absent from the Act itself. Some observers argue that this means the new Act enables compulsory acquisition by other private individuals, without guaranteeing any public benefit in return (Sherry, 2015). This contravenes the principles that have guided compulsory acquisition in the Australian context – that the benefit does not accrue to a private party, and that the acquisition achieve a public purpose. The Act does require some judicial oversight by the Land and Environment Court, but only to ensure that "the terms of the settlement [received by owners] under the plan are just and equitable in all the circumstances". It remains to be seen how the Court will interpret this obligation, and whether there is any capacity to include broader public purpose considerations in the determination of what is "just and equitable in all the circumstances". It is worth noting, however, that the Minister responsible for the legislative change described the Court's powers as providing "definitive protection", and that "there is no chance vulnerable people will be forced into worse situations with the court given the power to look at all the circumstances" (Robertson, 2016). This has not proven to be the case in other jurisdictions, however; for example, in Singapore the courts only intervene if proper procedure has not been followed, or dissenting owners do not receive appropriate compensation. The courts have declined to change the decision on any other hardship grounds (Christudason, 2009; see also Chapter 10).

Notably, there is existing legislation in NSW that would allow the government to involve itself in the renewal of at least some strata titled buildings, and to ensure a public benefit was achieved. Of particular interest is the *Growth Centres (Development Corporations) Act 1974* (NSW) which allows a government development corporation to acquire land in, or adjacent to, any areas defined as growth centres. Areas are defined through the *State Environment Planning Policy (Urban Renewal) 2010* (Urban Renewal SEPP), which aims to facilitate redevelopment of urban renewal precincts to achieve metropolitan planning objectives. Currently, these growth centres exist in Granville and Waterloo, but the list could be expanded by amending the SEPP. This would allow government-led densification in areas where it has already been identified as a strategically beneficial outcome.[5] Instead, the government has chosen an approach that allows renewal to be directed by private property owners acting in their interests only.

Looking at these issues in aggregate, it seems reasonable to view the new strata termination process as a form of de facto compulsory acquisition. Furthermore, it is a form of compulsory acquisition that does not meet established standards of producing a public benefit and eschewing private gain, and is therefore more likely to produce inequitable outcomes than compulsory acquisition laws, particularly for vulnerable residents. This is not the only concern raised about the new process, however, as the next section explores.

Other social concerns regarding strata termination

Given the similarities with compulsory acquisition, a process that often creates public controversy, it is surprising that the introduction of the new strata renewal process did not generate more heated public debate (see Thomson, 2015; Robertson, 2016 for some of the few detailed press articles). One explanation is that many Australians do not fully understand the nature of strata title, and therefore do not appreciate the significance of this change (Easthope, 2015). Another possible explanation is the high rate of investor ownership in strata, which means that over 55 per cent of strata owners would lose an asset (but not their home) if their scheme is terminated.

Of course, there is a legal reality regarding property ownership which means that there are already many ways people can be dispossessed of their homes (for example, a mortgagee enforcing a power of sale or foreclosing under a mortgage). This situation applies equally to houses as it does to strata titled units, regardless of any legislative change. However, the social realities of ownership are different, as it is commonly understood that purchasing a home means you can keep it as long as you desire, reinforcing a notion of 'ontological security' (Hulse *et al.*, 2011). This notion has been promoted in relation to strata title in much the same way as with freestanding houses (Troy *et al.*, 2017). As such, the loss of security will likely hit many strata owners very hard, as one interviewee from a consumer peak body noted: "I think the loss of security of tenure is the primary one. These people bought into their homes on the understanding that it would be theirs for as long as they wanted it."

This understanding is reinforced in many ways through current approaches to housing policy, which maintain home ownership as almost sacrosanct and deeply embedded within the Australian model of citizenship (Kemeny, 1983; Troy, 2012). This broad political consensus on home ownership and citizenship means that owning one's home is imbued with more than the bundle of rights afforded under common law and legislation.

Even beyond the (now significant) differences in legal rights between strata and non-strata titled properties, however, there are also differences in ownership patterns that will likely exacerbate the impact of the new strata renewal process. Two factors distinguish the community of strata owners and residents from those in non-strata properties:

- median household incomes are lower for apartment residents than for people living in detached houses; and
- the owners are more likely to be investors, rather than owner-occupiers (ABS, 2017).

These dynamics reinforce the risk that strata termination will disproportionately affect more vulnerable residents, as they disproportionately live in strata. They are also likely to have different interests to investor owners, who nevertheless have an equal say to owner occupiers in how this process plays out.

While many strata residents who participated in our research were unaware of the proposed changes to strata termination, they were well aware of the distinctive features of strata living more generally (on these features see Easthope and Randolph, 2009; Sherry, 2013). These factors clearly shaped their responses to the proposed legislative change, which were nuanced, and concerned with both social and legal implications. The next section provides a brief overview of some key elements of their responses.

Research participant responses

Despite the issues outlined above, our research found that strata stakeholders and residents have mixed views on the idea of lowering the termination threshold below 100 per cent. They identified various factors to consider when deciding whether a forced sale might be justified. These included an owner's motivation for opposing a sale, what public benefits might flow from the sale, and how a forced sale might be managed.

Is a reduced threshold positive, and on what basis is it justified?

To begin with, the survey of strata residents showed that respondents were mixed in their opinion of the proposed changes (see Figure 11.1). While 49 per cent of renters and 43 per cent of owner-occupiers supported reducing the threshold, 58 per cent of investor-owners supported changes (albeit only a small sample size).

These mixed reactions were confirmed in both the community workshops and stakeholder interviews, where views on a reduced threshold were regularly qualified by various caveats. When participants were asked what they assumed the purpose of the legislative change to be, most thought it was primarily to address 'hold out' situations, where a single resident prevents a strata termination from proceeding. This was considered a legitimate issue for legislative action, but participants were keen to stress that hold outs can take two forms. The first is a person holding out simply to achieve a higher price, as the prospective buyer is left with no choice if they wish to obtain ownership of the whole strata block. As an interviewee from a property industry peak body pointed out, the second is a person who does not want to leave their home for personal reasons, or lacks the financial capacity to move: "Two people are holding out. One, the little old lady, who doesn't know what to do. The other an investor, who just thinks by holding out to the last person, he'll get more than everyone else."

With one exception, all interviewees thought that it was acceptable for the 'old lady' to hold out where she would otherwise be left without a unit, or be forced out of the area. The elderly were seen as among the most vulnerable both in terms of capacity to cope with change (financially, emotionally, physically)

Q. The NSW Government has proposed to change the legislation to enable a strata scheme to be cancelled to allow the building to be demolished if a majority (possibly 75%) of owners agree. Do you agree with this proposal? (n = 307 renter; 907 owner-occupier; 35 investor-owner)

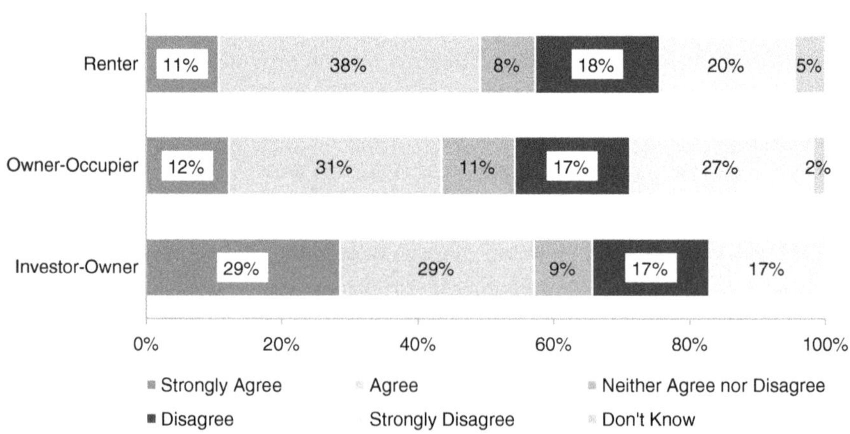

Figure 11.1 Community response to change in voting proportions.

and the capacity to navigate the termination process successfully. But there was little support for those holding out purely for speculative gain, as this was considered unfair to other owners. As one industry workshop participant explained:

> You will get some people who will not agree with anything unless they get $2 million for a $100,000 property. So it's to stop that, it seems to me – that unreasonable blocking. Because it might not be 25 per cent, it might just be one person who's holding out.

While this distinction simplifies what may actually be a diverse range of justifications for resisting a sale, the tendency to draw this distinction highlights the need for reasons for 'holding out' to be taken into consideration. In other words, beyond any legal rights sit moral and social considerations, which respondents felt were also relevant to determining the legitimacy of termination proceedings.

The distinction here is also drawn between those who invested in property as a financial product, and those who occupied property as a home. At the same time, however, both sets of actors were seen as legitimate. This tension between strata as housing with use value and as an investment with exchange value was explicitly recognised by some interviewees (see also Troy *et al.*, 2017). As an interviewee from a consumer peak body explained:

> You should not treat somebody's roof over their head as an investment thing primarily. That's why it should not be possible to simply force somebody to sell because there's a buck in it for a majority of the other owners.

With one exception, all interviewees acknowledged that emotional, cultural and political aspects of housing played an important and legitimate role in the process of urban renewal, and should be taken into consideration when determining outcomes. There was, however, significant disagreement about the extent to which such intangible values should affect the outcomes. With the exception of consumer bodies representing disadvantaged groups, most participants did not believe this should be a complete barrier to termination occurring.

During community workshops, these distinctions translated into a suggestion that a fixed threshold was not the most appropriate way to address the issue. These participants thought that any change to the voting threshold should instead be designed to deal specifically with the 'hold out' scenario. To demonstrate how this might work in practice, we mapped the distribution of strata schemes in Sydney by size, and the variation in voting proportions needed to address single hold outs (Figure 11.2).

This analysis demonstrates that if the aim of the Act was to prevent a single 'hold out' from manipulating the termination process for unfair personal gain, the voting threshold could be increased and tiered according to strata scheme size, starting at a minimum 80 per cent.

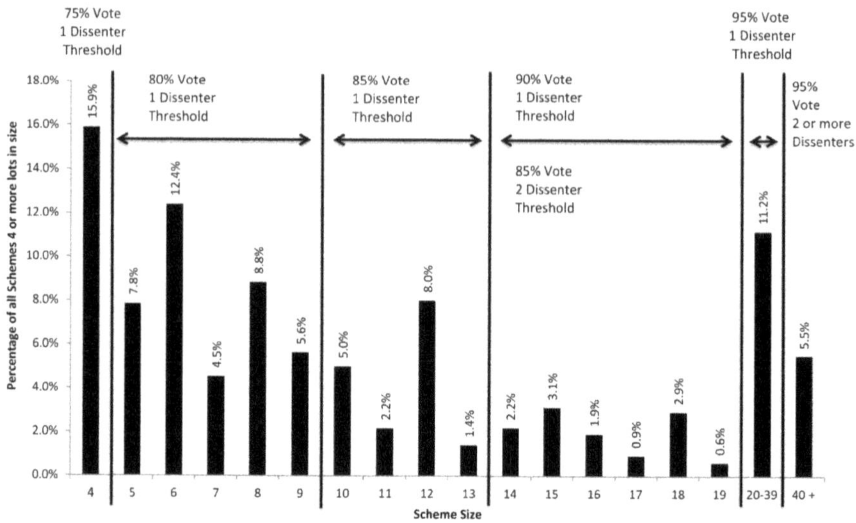

Figure 11.2 Voting proportions designed to address a single 'hold out'.

What other factors affect whether a forced sale is legitimate?

Some owners also felt a public benefit (including better quality buildings, or new infrastructure to support higher density development and population growth) would make them more amenable to a sale. These results indicate that questions of public good are a factor in how some owners choose to exercise their private property rights. Furthermore, few believed property rights should be absolute, or that housing's exchange value as a financial asset should always prevail over its use value as a social good.

To unpack these nuances in more detail, the survey asked respondents about the impact of different factors on whether they would support a proposal to redevelop their building.

Building quality and safety

If a building was in poor or unsafe condition then termination was largely supported, with renters more supportive than owner-occupiers.

Significance of broader planning objectives

An appreciation of the wider significance of strata renewal in terms of neighbourhood impact was also clear among survey respondents, with resident owners overwhelmingly claiming this would be an important influence on their decision to support a termination proposal (see Figure 11.5). Perhaps surprisingly,

Q. In what circumstances would you agree or disagree to redevelop a scheme without the support of all (100%) owners?

(i) The building/scheme is in poor physical condition (n = 304 renter; 888 owner-occupier; 36 investor-owner)

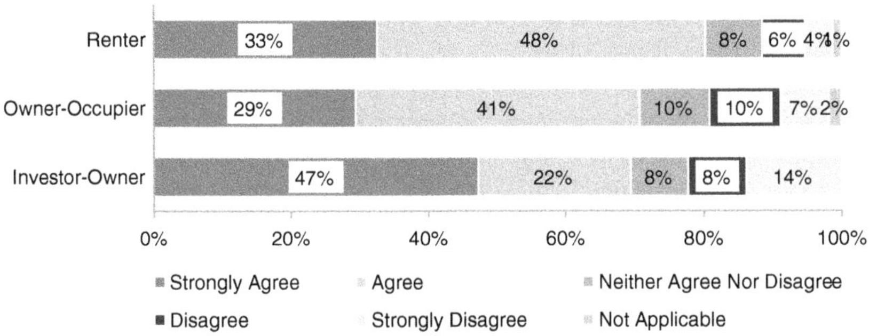

Figure 11.3 Community attitudes towards renewal to address poor building condition.

(ii) The building does not meet current building safety standards (n = 305 renter; 884 owner-occupier; 34 investor-owner)

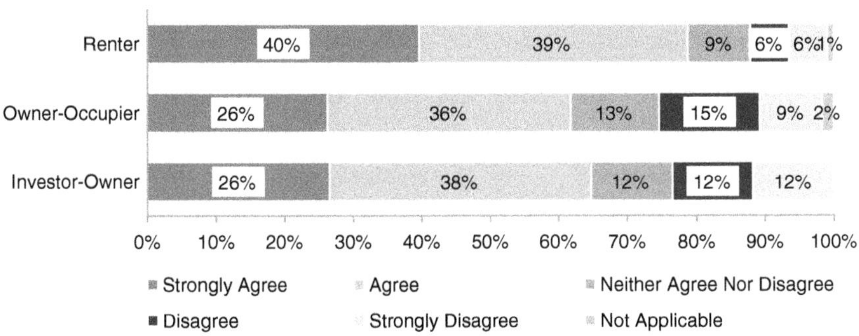

Figure 11.4 Community attitudes towards renewal to address safety standards.

Q. If it came to a vote to redevelop your scheme, how important would the following issues be in deciding whether you would support a proposal? Impact of the redevelopment on the local neighbourhood (n= 887 owner-occupier; 36 investor-owner)

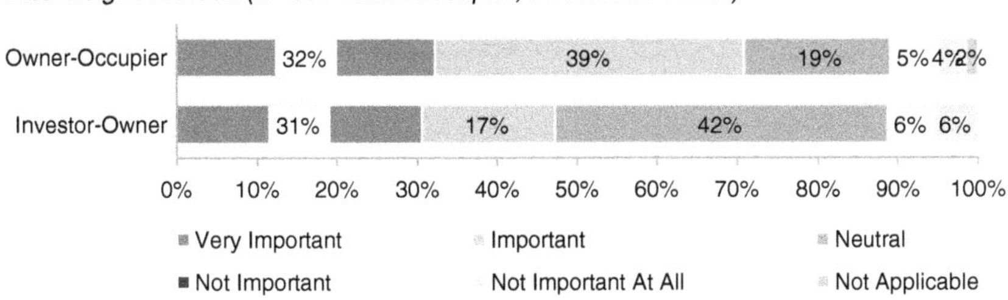

Figure 11.5 Community attitudes on importance of impact on local neighbourhood.

this view was also widely supported among investor-owners, with almost half agreeing that neighbourhood impacts would be of some significance.

On the other hand, it was clear that owners were not convinced that they should be expected to sell simply because the government is seeking to increase housing supply. When asked whether the need for more housing in their area would be an important factor in deciding whether to sell, the majority were at best ambivalent (see Figure 11.6).

A similar response was recorded for both renters and owner-occupiers to the suggestion that their scheme could be rezoned for higher density to support redevelopment – only 23 per cent supported this proposition, while well over half opposed it (see Figure 11.7). There is a clear resistance among strata residents to the idea that they should make way simply to achieve higher density renewal. However, investor-owners were much more willing to contemplate termination for this reason.

Q. If it came to a vote to redevelop your scheme, how important would the following issues be in deciding whether you would support a proposal? The need for more housing in the area (n = 880 owner-occupier; 36 investor-owner)

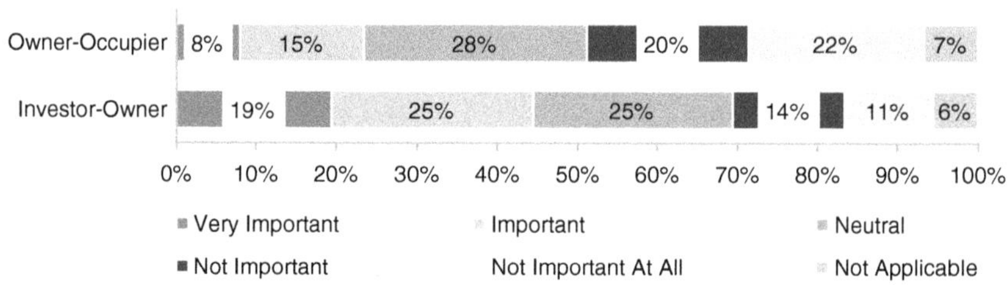

Figure 11.6 Community attitudes on importance of need for more housing in area.

Q. If your local area was rezoned for higher densities and a larger building could be built on the land on which your current block is located, how willing would you be to support the redevelopment of your strata scheme? (n = 307 renter; 904 owner-occupier; 36 investor-owner)

Figure 11.7 Community attitudes on renewal for rezoning to higher density.

Fairness and oversight of the process

Overall, these responses indicate strata residents have a nuanced view of the need to balance private property rights and the ontological security of home on one hand, and the broader public benefits of urban renewal on the other. To ensure these outcomes are achieved, owner-occupiers and investor-owners showed significant support for formal court oversight and the capacity to contest termination decisions on multiple grounds, some of which potentially sit outside procedural aspects of renewal. Over 85 per cent of owners supported oversight by the Land and Environment Court.

One of the strongest themes to emerge from interviews and workshops was the desire for a process that was considered fair and transparent. In the survey, 93 per cent of all owner-occupiers and 91 per cent of all investor-owners rated fairness and oversight as either very important or important when it came to making a decision on a renewal proposal (see Figure 11.8).

Strong support for equitable outcomes is evident, both in the concerns raised about vulnerable residents, and the disapproval voiced about owners using the system to achieve disproportionate personal gain. And while better neighbourhood outcomes are considered a legitimate reason for strata termination, there is far less support for termination simply to achieve greater density. Together, these responses offer useful insights into the likely public response as strata terminations occur across Sydney in coming years.

Conclusion

As this chapter has shown, a comparison between NSW's new strata termination model and its existing compulsory acquisition regime highlights some key similarities, including:

- in both cases, the power to dispossess is granted by way of legislation, not through any constitutional power;

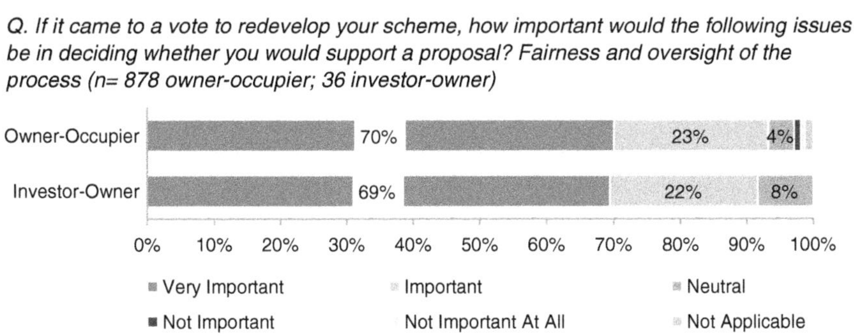

Figure 11.8 Importance of fairness and oversight of renewal process.

- both processes require the payment of compensation, based on the principle of 'just terms'; and
- there is no restriction in either case on the locations where termination/ acquisition can occur.

At the same time, however, two key differences emerge:

- the strata termination procedure affords more limited protections against unjust dispossession than the compulsory acquisition procedure; and
- there is no requirement to demonstrate a public benefit before strata termination can occur. While this is also technically true for compulsory acquisition by state and municipal bodies, in practice governments have only used this power in cases where a public benefit is evident.

These two differences are significant. As the practice of removing the legal requirement for unanimous approval to terminate a strata scheme gains popularity around the world, it is important that the risks associated with this major shift in property rights are acknowledged and addressed. While strategic strata renewal can undoubtedly produce public benefits, the NSW experience demonstrates that new legislation to facilitate non-unanimous strata termination will not necessarily require that such benefits are achieved. For this reason, it is important that the implications of this change are made clear, particularly the way it diminishes the property rights of strata title owners relative to the rights enjoyed by other freehold owners. As a result, the strata termination process contains fewer protections against unjust dispossession than the process used by local municipalities to compulsorily acquire land, which is already highly controversial.

This distinction seems hard to justify, particularly given the current political imperative to produce increasingly dense and compact cities, with growing numbers of residents living in strata. Strata title has always been different from other forms of freehold title, but its dramatic growth in urban areas indicates it is now the only form of property ownership available to many Australians. This is particularly true for lower-income and vulnerable residents. The new strata termination procedure does include a safeguard requiring that the terms of the strata termination and renewal plan be "just and equitable", but it will be up to the courts to determine how this will be interpreted, and it may be on purely financial grounds. In that case, it remains likely that lower-income and vulnerable residents will bear most of the negative impact of forced strata terminations, for the reasons outlined in the second section.

The introduction of the new strata legislation did not garner as much public attention as might have been anticipated, so it remains to be seen what the public reaction will be once forced strata terminations start occurring. There appears to be significant market demand in Sydney for large, high density redevelopment sites, as the recent 'superlot' sales have shown (these sales involve multiple neighbouring property owners banding together to sell their combined properties to a developer for significantly inflated prices – see e.g. Cummins, 2017). This suggests the market drivers are in place for strata terminations to proceed at pace. In these circumstances, it remains to be seen whether the financial incentives will ultimately override the concerns many strata owners have expressed about the new termination process.

Nonetheless, it is significant that our research has shown that many owners are cognisant of strata requiring a careful balancing of personal and collective rights, and claim to be as much concerned about equitable treatment as about maximising their financial benefit. While much public debate focuses on the economics of property ownership, these findings suggest that social and emotional ideas of 'home' also remain important considerations. At the same time, however, respondents were quite particular about identifying which public benefits they consider to be legitimate reasons for overriding the personal property rights of strata owners. As such, those governments considering similar legislative changes would be well advised to design the process in a way that ensures it enhances the public amenity of our urban areas, not simply the economic standing of some strata owners.

Notes

1 The existing legislation did actually allow scheme termination without unanimous owner consent via an application to the Supreme Court. This provision was rarely utilised, however, and as of April 2012, only 826 NSW strata schemes had ever been terminated by either method (Easthope *et al.*, 2013).
2 Or 90 per cent for buildings under ten years old.
3 This can be reduced to 80 per cent by the government.
4 Refers to the discussion papers and positioning papers published by NSW Government (2012; 2013).
5 Similarly, the *Housing Act 2001* (NSW) empowers the Land and Housing Corporation to compulsorily acquire, develop and dispose of land and housing (ss. 18, 22).

References

ABS (Australian Bureau of Statistics) (2017) *Apartment Living*. Stories from the Census, 2016. Available at: www.abs.gov.au/ausstats/abs@.nsf/Lookup/by%20Subject/2071.0~2016~Main%20Features~Apartment%20Living~20 (accessed 10 April 2018).

Bunker R, Crommelin L, Troy L, Easthope H, Pinnegar S and Randolph B (2017) Managing the transition to a more compact city in Australia, *International Planning Studies*, 22(4), 1–16.

Christudason A (2009) Property rights: Achieving a fine balance in collective sales of strata developments in Singapore, *International Journal of Law in the Built Environment*, 1(1), 26–41.

Commercial and Property Law Research Centre and QUT Law (2014) *Queensland Government Property Law Review Options Paper Body Corporate Governance Issues: By-Laws, Debt Recovery and Scheme Termination*. Available at: www.justice.qld.gov.au/__data/assets/pdf_file/0003/334758/Property-law-review-Body-Corporate-Governance-Options-Paper-1.pdf (accessed 10 April 2018).

Cummins C (2017) Super-lot sales show no sign of slowing, *Sydney Morning Herald*, 27 September. Available at: www.smh.com.au/business/property/super-lot-sales-show-no-sign-of-slowing-20170926-gyozm2.html (accessed 10 April 2018).

Easthope H (2015) The role of retirees in residential 'private governments', *Journal of Urban Affairs*, 37(3), 311–326.

Easthope H and Randolph B (2009) Governing the compact city: The challenges of apartment living in Sydney, Australia, *Housing Studies*, 24(2), 243–259.

Easthope H, Hudson S and Randolph B (2013) Urban renewal and strata scheme termination: Balancing communal management and individual property rights, *Environment and Planning* A, 45(6), 1421–1435.

Garnett N S (2006) The neglected political economy of eminent domain, *Michigan Law Review*, 105(1), 101–150.

Hulse K, Milligan V and Easthope H (2011) *Secure Occupancy in Rental Housing: Conceptual Foundations and Comparative Perspectives*. AHURI Final Report no. 170, Australian Housing and Urban Research Institute: Melbourne. Available at: www.ahuri.edu.au/research/final-reports/170 (accessed 10 April 2018).

Kemeny J (1983) *The Great Australian Nightmare: A Critique of the Home-ownership Ideology*. Melbourne: Georgian House.

Landgate (2014) *Strata Titles Act Reform: Consultation Paper*. Perth: Landgate.

NSW Government (2012) *Making NSW No. 1 Again: Shaping Future Communities: Strata and Community Title Law Reform Discussion Paper*. Sydney: NSW Fair Trading.

NSW Government (2013) *Strata Title Law Reform: Strata and Community Title Law Reform Position Paper*. Sydney: NSW Fair Trading.

NSW Parliament (2015) *Second Reading Speech: Strata Schemes Development Bill 2015*. Available at: www.parliament.nsw.gov.au/Hansard/Pages/HansardResult.aspx#/docid/HANSARD-1323879322-73682/link/47 (accessed 10 April 2018).

Pocock M (2014) Compulsory acquisition, public benefits and large scale private sector developments: Can Australia learn from the United Kingdom? *Local Government Law Journal*, 19(3), 129–141.

Pocock M (2015) Orgies of seizure and violence: Compulsory acquisition and private sector development – lessons for Australia, *Local Government Law Journal*, 20(1), 27–40.

Property Council (2015) NSW strata laws get a makeover, *Press Release*, 3 November. Available at: www.propertycouncil.com.au/Web/Content/News/National/2015/NSW_strata_laws_get_a_makeover.aspx (accessed 10 April 2018).

Robertson J (2016) Strata dispute heats up over majority rules, *Sydney Morning Herald*, 30 May. Available at: www.smh.com.au/nsw/strata-dispute-heats-up-over-majority-rules-20160529-gp6n7e.html (accessed 10 April 2018).

Russell D (2014) *Review of the Land Acquisition (Just Terms Compensation) Act 1991*. Report prepared for the Minister for Finance and Services. Sydney: NSW Government.

Searle G (2004) The limits to urban consolidation, *Australian Planner*, 41(1), 42–48.

Sherry C (2013) Lessons in personal freedom and functional land markets: What strata and community title can learn from traditional doctrine of property, *UNSW Law Review*, 36(1), 280–315.

Sherry C (2015) Strata law overhaul a step too far, *Sydney Morning Herald*, 23 August. Available at: www.smh.com.au/comment/strata-law-overhaul-a-step-too-far-20150823-gj5mz5.html (accessed 10 April 2018).

Sherry C (2017) *Strata Title Property Rights: Private Governance of Multi-owned Properties*. Abingdon, Oxon: Routledge.

Thomson S (2015) There's a huge change coming for the two million people living in apartments in NSW, *Business Insider*, 29 October. Available at: www.businessinsider.com.au/theres-a-huge-change-coming-for-the-two-million-people-living-in-apartments-in-new-south-wales-2015-10 (accessed 10 April 2018).
Troy P (2012) *Accommodating Australians: Commonwealth Government Involvement in Housing*. Annandale: Federation Press.
Troy L, Randolph B, Crommelin L, Easthope H and Pinnegar S (2015a) *Renewing the Compact City: Economically Viable and Socially Sustainable Approaches to Urban Redevelopment*. Sydney: City Futures Research Centre, UNSW.
Troy L, Easthope H, Randolph B and Pinnegar S (2015b) *Renewing the Compact City – Interim Report*. Sydney: City Futures Research Centre, UNSW.
Troy L, Easthope H, Randolph B and Pinnegar S (2017) 'It depends what you mean by the term rights': Strata termination and housing rights, *Housing Studies*, 32(1), 1–16.

12 Compulsory acquisition of private property rights for densification in Australia

John Sheehan and Charlie Glinka

Introduction

There are a number of methods of compulsory land acquisition when the need for urban consolidation or densification confronts government. The method with which acquisition is undertaken "is a key variable in the success of urban renewal programmes" (Erik, 2008, 70). The choice of method can depend on the behaviour of owners in possession of nominated land, and a distinction needs to be made between active and passive landowners (Adams, 1994, 103). Active landowners will develop their land to increase its value or will make their land available for development by others. Passive landowners "take no particular steps to market or develop their land and therefore contribute little or nothing to the development process. They refuse to sell the land or are only willing to sell under restrictive terms or conditions" (Adams, 1994, 104). There is no way of ascertaining whether landowners who refuse to sell are strategically holding out and hence active, or merely asserting their private property rights (Kelly, 2005, 1).

Concomitant with compulsory acquisition of private property rights is the issue of entitlement to compensation. Australian property law has revealed with the belated recognition of native title by the High Court in the 1992 decision *Mabo v. Queensland (No. 2) (1992) 175 CLR 1* that the ambit of prospective compensation for the commutation of a whole raft of private property rights (Indigenous or non-Indigenous) by government has widened significantly. Difficulties in descriptors of specific rights are not unexpected given the increasing complexity of private property rights, and hence the compensation assessment of such losses is being forced into often wholly unfamiliar legal and practice territory. Yet, linguistic obfuscation remains a significant barrier in attempts to unravel the particular complexity of specific property rights with such endeavours often thwarted by connotation integral to those tenures easily recognised as property, and hence readily compensable.

In Australia, the compulsory acquisition of any private property rights for urban consolidation or densification, or indeed any compulsory acquisition, nevertheless has its roots in a complex amalgam of custom and law, Ely (1998, 10) observing

> English common law provided the legal foundation for property ownership in the colonies. Common law was customary law, deriving its authority from long-established usage. Royal courts in England fashioned the common law into a body of rules that defined and protected property rights.

Conceptually, private property rights and the concept of value necessarily emerged as the twin *leit motifs* of English common law, and its colonial American and Australian progeny. Importantly, private rights in property arose with the emergence of the early modern state wherein according to Anderson (1979, 429) "the age in which 'Absolutist' public authority was imposed was also simultaneously the age in which 'absolute' private property was progressively consolidated".

However, private property rights are all held in Australia at the pleasure of the Crown, which has the ability to compulsorily acquire those private rights subject to some safeguards (and even guarantees in some jurisdictional circumstances).

The practice of compulsory acquisition of private property rights

In the Australian State of New South Wales (NSW)[1] how compulsory acquisition of private property rights by government for urban consolidation and densification is practised and subsequently compensation assessed arguably reveals an underlying flaw in the social narrative between the citizenry and the NSW government. Broadly, the practice of compulsory acquisition of private property rights for a lawful purpose is impacted by international law, such as the *United Nations Declaration of Human Rights* (United Nations, 1948) at *Article 17* which states that "No one shall be arbitrarily deprived of his property." More recently, in 2008 the UNFAO published a guide to "good practice" entitled Compulsory Acquisition of Land and Compensation (UNFAO, 2008), which establishes various criteria pertaining to compulsory acquisition processes and compensation. However, the Australian Law Reform Commission (ALRC) has soberly noted international instruments such as those mentioned above are not part of domestic law until recognised in statutes, subject to the caveat: "where a statute is ambiguous, courts will generally favour a construction that accords with Australia's international obligations" (ALRC, 2015, 487).

Irrespective, the compulsory acquisition by government of private property rights for whatever lawful purpose presents a challenge to most holders who are "completely unfamiliar" with legislation and other processes involved in compulsory acquisition and significantly at a disadvantage to the "highly resourced, extensively experienced" acquiring authority (Shannon, 2012, 9). Yet, redolent of the *Fifth Amendment* of the *United States Constitution* which was drafted as "the primary means of protecting private property from the excesses of government" (Gardner, 1997, 542), in the Australian *Constitution* there is also the re-occurring theme of protecting private property from such excesses through a guarantee of *just terms* compensation (Irving, 1999, 74, 96; De Soyza, 2017, 455).

However, in NSW the Commonwealth Constitutional guarantee of just terms compensation does not apply to government compulsory acquisition of private property rights, although in 2009 in *R&R Fazzolari Pty Limited v Parramatta City Council; Mac's Pty Limited v Parramatta City Council* HCA 12 (2 April) ([2009] 2009) the High Court observed: "in accordance with established principles of statutory interpretation the preferable construction is that which authorizes the least interference with private property rights" (*Fazzolari* at [5]).

In *Fazzolari*, Parramatta City Council intended to acquire private land that would subsequently be sold and redeveloped to form part of land to be called the 'Civic Place'. To compulsorily acquire the private land, Council issued Proposed Acquisition Notices (PANs) to various private landowners within the proposed 'Civic Place', one being Fazzolari. The intention was that once the private lands were acquired, a AUS$1.6 billion redevelopment would be undertaken by means of a Public Private Partnership between Parramatta City Council and a developer, GROCON. The 'Civic Place' development agreement proposed a massive increase in density within the Parramatta CBD comprising in part "a residential tower of 31 storeys, known as Park Apartments", "a 40 storey commercial office tower, known as the Atria" with a "level of retail and four levels of carparking beneath the surface" (*Fazzolari* at [8]). The development agreement required Council to transfer some of the acquired land to GROCON in return for significant financial payments, "12,000 square metres of [new] public open space in the Parramatta CBD together with various other community facilities" (*Fazzolari* at [15]).

In the Land and Environment Court, Fazzolari argued that this would constitute a 'resale' for the purposes of *s 188(1)* of the *Local Government Act 1993*, an unlawful action. The Court found in favour of Fazzolari and held the compulsory acquisitions were unlawful. Council subsequently appealed to the NSW Court of Appeal where the decision was reversed. Finally, Fazzolari was granted special leave to appeal in the High Court and the original decision of the Land and Environment Court was reinstated.

It was held the power of local government (such as Parramatta City Council) to compulsorily acquire private land lies in *s 186 Local Government Act 1993*, but constrained by *s 188*:

> (1) A Council may not acquire land under this part by compulsory process without the approval of the owner of the land if it is being acquired for the purpose of re-sale.
>
> (2) However, the owner's approval is not required if:
>
> > (a) The land forms part of, or adjoins or lies in the vicinity of, other land acquired at the same time under this Part for a purpose other than the purpose of re-sale.

The first issue was whether the acquisition and then transfer to GROCON would be a 're-sale' within the meaning in *s 188(1)*. The High Court held that the acquired land was to be transferred in exchange for money and other consideration, rendering it a 're-sale'. Further, it was enough to attract the constraint in *s 188(1)* that one of the purposes of the proposed acquisition was for re-sale, even if it could be argued that the dominant purpose was for redevelopment.

The second issue was whether the surrounding streets, Darcy and Church, being acquired at the same time meant that Council could also acquire Fazzolari's land under *s 188(2)(a)*. Parramatta Council delayed the acquisition of the streets so that it would coincide with the acquisition of Fazzolari's land, in order to rely on *s 188(2)(a)* in case the 're-sale' argument was not allowed. However, the High Court found that in this instance the Council's ability to acquire its own roads (Darcy and Church Streets) was not found in the *Local Government Act 1993*, but rather relied on *s 7B Land Acquisition (Just Terms Compensation) Act 1991*: "An authority of the State that is authorized by law to acquire land by compulsory process in accordance with this Act may so acquire the land even if the land is vested in the authority itself."

The fact Parramatta City Council was acquiring land for the purpose of transfer for money and other consideration to permit (arguably) urban consolidation and densification for a profit invalidated the compulsory acquisition of the private property rights of Fazzolari. The High Court was unwilling to interpret statutory regulations in a manner that might unduly infringe on private property rights, and accorded with longstanding common law presumptions that Parliament's intention is not to interfere with fundamental rights unless the precise wording of the statute says so.

Commentary on the presumption in 1765 from William Blackstone in Commentaries on the Laws of England (1765, 135) asserts that the common law would not allow for the 'least violation' of private property. In 1904, Griffith CJ expressed in *Clissold v Perry (1904) 1 CLR 363* that the construction of statutes are not to be interpreted to interfere with vested interests, unless the intention is obvious. Most recently in *Fazzolari*, French CJ stated that:

> … [a]s a practical matter … where a statute is capable of more than one construction, that construction will be chosen which interferes least with private property rights … Constructional choices are open, so that they do not encroach upon fundamental rights and freedoms at common law.
>
> (*Fazzolari* at [43])

In spite of such deeply rooted common law principles and the High Court's decision in *Fazzolari*, in 2009 the NSW Parliament amended *s 7 Land Acquisition (Just Terms Compensation) Act 1991* to permit compulsory acquisition of land by Councils for re-sale in *s 188(2)*.

Kirby J. explained in *Durham Holdings Pty Ltd v the State of NSW (2001) 205 CLR 399 at [72]* that an extreme departure from fundamental rights may be challenged on a constitutional basis. His Honour opined that the role of the Constitution for judicial protection in the face of legislation is 'substantial'.

Under what circumstances can compulsory acquisition of private property rights occur to permit urban consolidation and densification?

Soberly it is recalled the High Court in *New South Wales v Commonwealth* (1915) 20 CLR 54 (known as the Wheat Case) confirmed State Parliaments retain "sovereignty to make laws for the compulsory acquisition for private property without payment of compensation" (Raff, 2002, 40) should a State so decide. Indeed, there is a history of reduced compensation for specific projects in NSW such as the Eastern Suburbs Railway where compensation was limited to 27 February 1967 in the *City and Suburban Electric Railways Act 1915–1967* irrespective of the settlement of the compensation claim many years later (Sheehan, 2010, 108).

That the previously mentioned 2009 amendment of *s 7 Land Acquisition (Just Terms Compensation) Act 1991* constitutes an extreme derogation from fundamental rights remains an open question, and the NSW government continues to rely upon *Art.9 Bill of Rights 1689 (Imp.)* which states "that the freedom of speech and debates or proceedings in Parliament ought not to be impeached or questioned in any courts or places out of Parliament".[2] This assertion is made in *Egan v Willis (1998) 195 CLR 424 at [22]*, and in the *NSW Legislative Assembly Practice, Procedure and Privilege* and the *New South Wales Legislative Council Practice*. Further, it was held in *Shaw v State Planning Authority of New South Wales (No. 2)(1972) 27 LGRA 94 at [10]* that the Proposed Acquisition Notice (PAN) which actually effects the acquisition under the *Land Acquisition (Just Terms Compensation) Act 1991* does not need to identify the specific purpose of the compulsory acquisition. The decision in *Shaw* followed the earlier High Court ruling in *Tinker Taylor Proprietary Limited v Commissioner for Main Roads (1960) 105 CLR 334 at 353 (per Kitto J, Dixon CJ agreeing)*.

Alternatively, whether the compulsory acquisition of private property rights is within power in the relevant legislation was considered to be "a right in point of justice" accruing to the holder of a private property right in *Jones v The Commonwealth (1936) 109 CLR 475 at 483 (per Dixon J)*. Such a right whether expressly or by implication in the interpretation of statutes necessitates there be sufficient particularity in the acquisition notice to enable the holder of a private property right to determine it is issued within power by the government. As mentioned earlier in *Fazzolari*, the High Court stated there is a presumption against an intention to interfere with private property rights (*Fazzolari at 618–619, [40]-[42]*). Further, it was held in *Minister for Public Works v Duggan (1951) 83 CLR 424 at 445* if the proposed compulsory acquisition is improper as to purpose, then the compulsory acquisition is invalid. Additionally, where government has been found to have acted for a substantial purpose which is unauthorised by statute and hence improper, the overall action has also been found to be invalid (*Thompson v Randwick Corporation (1950) 81 CLR 87 at 106; Samrein Pty. Ltd. v Metropolitan Water Sewerage & Drainage Board (1982) 412 ALR 467 at 468–469; Knuckey v Commissioner of Taxation (1998) 87 FCR 187 at 196)*.

Conclusion

Notwithstanding the decision in *Fazzolari*, some local government land use planners understandably supported the 2009 amendments to the *Land Acquisition (Just Terms Compensation) Act 1991*, legislative changes which have arguably "facilitated the transformation of the Parramatta City Centre through the Parramatta Square Development [and] are an example of urban renewal facilitation". When urban consolidation or densification involves site amalgamation it has been posited by planners that compensation should be "more than market value for land acquired for redevelopment. This, above market payment, should be reflective of a share of the uplift in value that will come from amalgamation" (Weekley *et al.*, 2017, 27).

From the above, three broad conclusions are inferred. First, cases such as *Fazzolari* invoke a variety of reactions within different disciplines such as law and planning. In particular, the complex nature of case law is reaffirmed reflecting the emerging complexity (and increasing monetary value) of private property rights when desirable actions such as urban consolidation and densification are confronted by government. Second, there needs to be improved communication between the acquiring authorities such as Councils and the dispossessed holders of private property rights, especially when their land compulsorily acquired is to be re-sold to a developer for urban consolidation and densification purposes. Compensation accruing to the dispossessed holder of such private property rights should arguably include a share of the resultant amalgamation value uplift, and necessarily should be part of the communication between the acquiring authority and the dispossessed holders. Third, swift legislative changes such as the 2009 amendments to the *Land Acquisition (Just Terms Compensation) Act 1991* in response to *Fazzolari* expose a disconnect between the citizenry and the NSW government. There is a need for a consensus as to what is fair compensation to enable compulsory acquisition of private property rights for urban consolidation and densification.

Acknowledgement

The authors acknowledge the invaluable assistance and guidance of Dr Spike Boydell, former Professor of the Built Environment and former Director of the Asia Pacific Centre for Complex Real Property Rights, Faculty of Design, Architecture and Building, University of Technology, Sydney (UTS).

Notes

1 NSW legislation is used in this chapter as an exemplar of the Australian legal milieu.
2 The text has been adjusted for modern spelling.

References

Adams D (1994) *Urban Planning and the Development Process*. Milton Park: Routledge.

Anderson P (1979) *Lineages of the Absolutist State*. London: Verso.

Australian Law Reform Commission (2015) *Traditional Rights and Freedoms – Encroachments by Commonwealth Laws: Final Report*, ALRC Report 129. Sydney: The Commission, December.

Blackstone (1765) *Commentaries on the Laws of England*, book 1, c 1.

De Soyza A (2017) Case Note and comment on Northern Territory v Griffiths, *Native Title News*, 12(7), 451–459.

Ely J W Jnr (1998) *The Guardian of Every Other Right: A Constitutional History of Property Rights*, 2nd ed. Bicentennial Essays on the Bill of Rights. New York: Oxford University Press.

Erik L (2008) Land assembly for urban transformation: The case of 's-Hertogenbosch in The Netherlands, *Land Use Policy*, 25(1), 69–80.

Gardner R C (1997) Taking the principle of just compensation abroad: Private property rights, national sovereignty, and the cost of environmental protection, *University of Cincinnati Law Review*, 65(2), 539–593.

Irving H (1999) *To Constitute a Nation: A Cultural History of Australia's Constitution*. Melbourne: Cambridge University Press.

Kelly D B (2005) *The "Public Use" Requirement in Eminent Domain Law: A Rational Based on Secret Purchases and Private Influence*. Fellows Discussion paper series, Cambridge: John M. Olin Centre for Law, Economics and Business, Harvard Law School.

New South Wales Legislative Council Practice.

NSW Legislative Assembly Practice, Procedure and Privilege.

Raff M J (2002) Planning law and compulsory acquisition in Australia. In: Kotaka T and Callies D L (eds) *Taking Land: Compulsory Purchase and Regulation in Asian-Pacific Countries*, 27–74. Honolulu: University of Hawai'i Press.

Shannon P (2012) Coal seam gas: Conduct and compensation agreements. Paper presented at the 50th Annual Queensland Law Society Symposium, 31 March.

Sheehan J (2010) Australia. In: Altermann, R (ed.) *Takings International: A Comparative Perspective on Land Use Regulations and Compensation Rights*, 107–118. Chicago: American Bar Association (Section of State and local Government Law).

UNFAO (Food and Agriculture Organization of the United Nations) (2008) *Compulsory Acquisition of Land and Compensation*. Rome: UNFAO.

United Nations (1948) *United Nations Declaration of Human Rights*.

Weekley G, Shannon C and Robinson K (2017) Overcoming the friction of site aggregation in the redevelopment of town centres, *New Planner*, June, 26–27.

13 Conclusion

Glen Searle

The use of compulsory property acquisition by governments to achieve public benefits has a long history. Emerging imperatives of sustainability have generated a new rationale for public intervention in areas such as urban development in order to produce environmental benefits that markets will not produce at adequate levels. This book has surveyed the use of compulsory acquisition to bring about better urban development in the form of densification, in a variety of nations across much of the globe.

While the benefits of densification might seem clear in most cases, the use of compulsory acquisition to gain these benefits is a contested and potentially fraught process. The protection of private property rights is generally regarded as a cornerstone of modern civic existence. Thus the threat to such rights from compulsory acquisition raises a range of issues relating to such matters as the nature and size of benefits that are produced in return, the adequacy of compensation to property owners, the extent of procedural justice and the influence of contextual factors such as constitutional decrees, the role of developers and other methods available for densification in each case. To this end, the book has attempted to understand how such issues are addressed in nine countries by posing the issues in terms of nine questions, as set out in Chapter 1. The following sections summarize the findings for the nine countries as they relate to each question/issue.

Justification for compulsory acquisition for densification

The justification given for compulsory acquisition for densification in theory and practice is central to understanding the use of such acquisition in each jurisdiction. The common element in each case is some form of public benefit that government intervention is required to bring about. The scope of expected public benefit varies significantly (as overviewed in the next section). There are also very varied circumstances in which compulsory acquisition to gain that benefit is regarded as justified: essentially, the costs of obtaining public benefits beyond monetary compensation to property owners are weighed to different extents across jurisdictions.

The justification for, and limits on the use of, compulsory acquisition in urban redevelopment are most fully set out in the statutes of European countries.

In Germany the use of expropriation has strict constitutional requirements. This limits it to specific cases which thus require much effort and that involve significant legal risks. The result is that compulsory acquisition is rarely used to achieve urban densification. The German legislation allows expropriation for certain defined planning purposes including densification on under-developed sites, and for the general good. The latter is only allowed in matters of 'vital and urgent' public interest. Otherwise expropriation is only admissible when its purpose cannot be reasonably achieved in other ways: it is only allowed as a last resort. The procedure for effecting compulsory acquisition is lengthy and very bureaucratic, in the interests of justice and fairness.

In the Netherlands, compulsory acquisition is only justified under strict conditions. Compulsory acquisition can only be used in the public interest, which is interpreted in fairly broad terms. It must be preceded by adequate negotiations to attempt voluntary purchase first. Property owners and all others affected must be financially compensated for all financial costs involved. Most compulsory acquisition cases result in a voluntary purchase, under the impending pressure on property owners of non-voluntary acquisition.

The justification for, and application of, compulsory acquisition in the UK is legally quite complex. In summary, however, compulsory acquisition must facilitate development, redevelopment or improvement of the land, but it must also improve or promote the economic, social or environmental well-being of the area concerned. Public authorities are required to negotiate with property owners to attempt to achieve a voluntary purchase before compulsory acquisition is used. As in Germany, such acquisition is only meant to be used as a last resort where there is a compelling public interest.

Spanish law allows a statutory land use plan to be the basis of compulsory acquisition. Two methods are legislated: expropriation and land readjustment (overviewed in later sections below). Expropriation is not often used, not least because of pro-owner emphases in Spanish jurisprudence. The land readjustment method is seen as offering a more reliable and financially feasible approach. The procedures for land readjustment are long and fairly complex. But the ability they give property owners to share in the value uplift from readjustment has meant that this version of compulsory acquisition has significant use in bringing about Spanish urban renewal and densification.

The compulsory acquisition rationale and process in Quebec (Canada) has some similarities with those in Spain. Quebec legislation allows local government to use compulsory acquisition of property to achieve redevelopment purposes of a statutory land use plan by assisting a developer to amalgamate land. The developer must already own, or have land owner acceptances to sell, at least two-thirds of the development area before compulsory acquisition is used.

In the US, the Constitution allows expropriation for 'public use'. The federal courts have interpreted this as 'public purpose' and under this view have allowed expropriation for economic development. However, the interpretation of what constitutes public use varies between the states. As Sullivan notes in Chapter 3, some states limit or exclude the use of expropriation for uses that are outside traditional government areas such as schools, libraries or road widening.

The justification and procedure for compulsory acquisition for urban redevelopment in Singapore were made explicit at the time the principal legislation was enacted. The legislation was devised to help achieve universal home ownership and a quality living environment by allowing sufficient land to be obtained for redevelopment. The theoretical justification for such acquisition, as Christudason observes in Chapter 10, has been that public necessity is greater than private concerns because the welfare of the public is paramount. The legislation requires compulsory acquisition to be for a 'public purpose', but the courts have ruled that what constitutes public purpose is at the government's discretion.

The justifications for compulsory acquisition in China are partly clear and partly opaque. The Link Policy's principal objective is to allow the expansion of urban areas while preserving the total area of farmland. An associated goal is to improve the living conditions of rural dwellers. Compulsory acquisition for urban redevelopment has a similar ostensible objective relating to the improvement of housing conditions. Underlying both rural and urban compulsory acquisitions, however, is the heavy reliance of local and regional governments on property sales as a source of finance. The potential for government profits to be made from purchasing private housing and obtaining land value uplift through new development on those sites (urban acquisitions) or selling land quotas to developers (rural acquisitions) has been a central driver of such compulsory purchases.

Australian compulsory acquisition legislation does not mandate any public benefits to be required. However, as noted in the next section, the acquisition powers under such legislation are generally used for purposes with public benefit, though rarely for densification. In the case study in Chapter 12, the courts ruled against the use of New South Wales compulsory acquisition legislation for enabling development of a high density centre because it benefited a private developer. New legislation was passed to allow acquisition in that case. The state's strata termination legislation, which allows compulsory acquisition of strata dwellings, also has no public benefit test, although the intention was to enable redevelopment of decrepit and dangerous buildings and serve wider planning objectives for urban consolidation.

Public benefits

The public benefits that are stated as the rationale for compulsory acquisition for densification vary considerably. Requirements range from a regulatory listing of different types of public benefit that can be applied, to mere unspecified 'public benefit' or, in the case of New South Wales strata title termination, no legally required public benefit at all. In general, systems in which the public benefit is not defined, and where there are also limited rights of challenge by those whose properties are being acquired, allow greater government discretion and freedom in using compulsory acquisition.

The most exactly regulated public benefit conditions are found in Germany. Compulsory acquisition is only permissible for the public good, as specified in certain permissible purposes relating to planning (including to 'assist' urban planning, and to use undeveloped or under-developed plots), or where it serves the 'general good'. The latter can only be invoked where there is a 'vital and urgent public interest'.

In the Netherlands, the interpretation of the public interest is fairly broad. The purposes of compulsory acquisition under Dutch legislation are set out under nine principles. The legislative principle of 'spatial development and housing' can be used to justify compulsory acquisition to implement land use plans, as long as compulsory acquisition is registered in a plan and it serves the public interest, public housing or the interest of spatial developments.

In Spain and Canada (Quebec), the public benefit is implicit through the purposes of public land use plans. In Quebec, municipalities can compulsorily acquire properties to achieve the objectives of a municipal plan. In Spain, such acquisition is permitted in order to bring about development permitted by the legally binding local plan (usually a municipal one).

In China, regulations for compulsory acquisition of urban properties for new housing contain eight situations that meet the public interest test, including economic development and redevelopment of areas with poor housing or insufficient infrastructure. In practice, however, as Zhang (Chapter 8) notes, the definition of public interest is vague and somewhat contradictory, and economic development is regarded as sufficient justification. China's Link Policy, involving compulsory acquisition of rural homesteads, has the specific stated aim of allowing more land to be provided for urban expansion without any decrease of total farmland.

Singapore's land acquisition arrangements have an equivalent stated aim, the provision of universal home ownership and a quality living environment, albeit stated by the nation's founding father, Lee Kwan Yew, soon after independence in the 1960s, and implicit since then. This aim has not been specifically incorporated into acquisition legislation. Land can be acquired compulsorily for any public purpose, benefit, interest and utility; or for any residential, commercial or industrial purpose. What constitutes public purpose etc. is not defined. This allows the government considerable discretion in using compulsory acquisition, with limited rights to challenge the basis of acquisition.

The United States Constitution's Fifth Amendment allows private property to be taken for 'public use'. As Sullivan points out in Chapter 3, there is no uniform definition of 'public use' across the 50 states, which have their own constitutions. The federal courts have conflated public use as 'public purpose', and under that definition have allowed (in *Kelo v. City of New London* [545 US 469 (2005)]) property foreclosure for economic development deemed as required for urban renewal by a public agency. This case reinforces previous court determinations in favour of public agencies' interpretation of public use.

The public benefit rationale in the United Kingdom for compulsory acquisition is arguably a little more determinate in many cases. Under the *Local*

Government Act 1972, acquisition can be for any purpose authorised by legislation or for 'the benefit, improvement or development of their area'. As Cox in Chapter 4 points out, this may rest on a widely drawn concept of public benefit, but also on the statutory functions of the public bodies exerting powers of compulsory acquisition. In the latter case, this can result in cases where the public body's powers are used for purposes that are 'tangential' to a direct public benefit.

In Australia, federal and state legislation allows the government to compulsorily acquire private property, but does not usually require a public benefit test although state governments usually restrict such acquisitions to projects with some public benefit. But Australian courts have not favoured compulsory acquisition if private entities will gain financially from the outcome, unlike the US in particular. In the case of the New South Wales *Strata Schemes Development Act 2015* which allowed termination of strata schemes with the approval of 75 per cent of strata title owners, the improvement of a dangerous or decrepit building was stated as a reason for the legislation, while the broader planning objective for densification of sites in high value areas was also a major driver. Nevertheless, there is no requirement in the Act that the strata termination serve some public purpose. At the same time, new development on the site of terminated schemes will have to conform to local planning controls, with their implicit incorporation of public benefit considerations.

Role of the development industry

There is wide variation in the role of the development industry in the process of compulsory acquisition for densification. This ranges from major involvement in the process, as in Spain, to merely acting on the development opportunities created by prior public acquisition.

As Chapter 7 outlines, the development industry in Spain has a central role facilitating compulsory acquisition for both urban expansion and urban renewal/densification via Spain's land readjustment procedures. Developers often initiate the land readjustment procedure, and obtain the support of municipalities and other public agencies to force readjustment on property owners who oppose it. The developers become the legal owners of expropriated land. Property owners themselves can join developers as voluntary partners. There are parallels here with the procedure in Quebec, Canada. The case study in Chapter 6 discusses how Montreal City Council has compulsory acquisition powers to acquire land to help developers amalgamate properties in order to achieve local plan outcomes; in that case, the local plan was also designed to suit the developer's own plan. In Australia, the strata title termination legislation was promoted and supported by the property industry, and it seems likely that developers will instigate the termination process in most cases.

In China, urban developers do not activate the Link Policy acquisition process, but require purchases of the land quotas produced by the policy to provide sites for (densified) residential development. In Chapter 9, Long Cheng suggests that the urgent demands of developers for land in urban areas have generated an acceleration of the process of rural homestead acquisition under the policy. In cities, compulsory acquisition of private properties by government for redevelopment has created profitable opportunities for developers, as discussed in Chapter 8.

The development industry's role in US compulsory acquisition (eminent domain) cases is essentially, as Sullivan notes in Chapter 3, that of an actor/stakeholder in the politics of such acquisition, in the same way as community associations and other interests. In that role, of course, pro-development and other business interests are able to liaise with city governments to propose economic development projects that might involve compulsory acquisition.

In the UK, the development industry may assist and facilitate the acquisition process, but it rarely directs that process. As Cox discusses in Chapter 4, the public acquiring authority makes an indemnity agreement with the development partner, which provides that the authority will exercise compulsory acquisition powers when appropriate and transfer the land acquired to the developer. In return, the developer agrees to pay some or all of the costs of the process plus an amount covering compensation due to the original property owners.

Compensation amounts

The compensation offered to owners of property compulsorily acquired is central to the issue of whether costs borne by them are outweighed by the benefits of resulting densification. The various jurisdictions covered in this book show wide variation in the extent of compensation and the degree of discretion involved.

The most extensive compensation provisions are made by European countries. In the Netherlands, affected property owners are financially compensated for all financial harm resulting from acquisition. This includes not only market value of the property acquired, but also decreases in the value of remaining property, and loss of income and other financial damage due to the compulsory purchase. The market value is based on a presumed sale in a free commercial market, which implies that value based on the future use of the land is included. Germany has similar provisions. Compensation must be provided for expropriation, either by money, land or other rights in lieu. The market value of land acquired is compensated, with the value calculated by independent committees of professional valuers using official ordinances or guidelines. In addition, compensation in Germany is provided for property loss of other kinds arising from expropriation, such as reductions in the value of remaining land (as in the Netherlands) or costs associated with removal activities required by expropriation.

In the UK, compensation is based on the principle that the amount payable to a person whose land is taken in the public interest should be equal to the 'true and actual loss' suffered by that person. This principle means the fair market value of the property concerned is paid to the owner, plus additional disturbance payments where a person is displaced from their residence (10 per cent of the market value up to a maximum limit), plus removal and related expenses. Related to this is the owner principle, whereby compensation for compulsory acquisition cannot include any increase or decrease in land value due to the scheme associated with the acquisition.

Spain's case is the most distinctive in terms of compensation. There, acquisition is normally carried out as part of a statutory land readjustment scheme. Compensation takes the form of building lots in completed schemes, distributed in proportion to the areas of original land holdings. These yield increased land values, with the amount of de facto compensation being the difference between new and old land values, minus costs of development (which are levied in proportion to the original size of properties). Where owners do not agree to become part of the redevelopment scheme, the development body (junta) created by the municipality compulsorily acquires their land. If they do not agree to pay their share of development costs, some of their property area is given to the junta in payment. Such owners receive serviced building lots at the end of the project, like those owners who participate voluntarily. Land readjustment regulations also allow public bodies to bypass the voluntary junta system and apply compulsory readjustment from the outset, but with the same ultimate financial benefits and costs to land owners.

Singapore's compensation provisions are also comprehensively defined, but with less favourable terms for land owners. Although changes since 2007 have improved compensation terms, particularly the introduction of compensation based on current open market value – which includes potential value realisable after redevelopment – other provisions limit payments to land owners. The most important is that compensation is to take into account any increase in the value of other property belonging to the same land owner likely to occur from the use to which the acquired land is put. As Christudason explains in Chapter 10, this is a form of value capture that reduces the cost of large public infrastructure projects. On the other hand, compensation is payable where proposed works reduce the value of retained land.

Compensation provisions for acquired property in China are quite detailed but legal protection for full compensation payments to be made seems to be lacking. For rural property expropriation, monetary compensation is payable for the residual value of the property, residents' social insurance, the value of agricultural outputs on the property, and relocation costs. For urban property expropriation, compensation is payable for residual property value and relocation costs. Land value is excluded in both cases, since property owners in China only have use rights on their properties. In addition, these due compensation payments are minimised in practice, arising from the incentives for this associated with the budgetary reliance of municipalities on land revenues. Zhang, in Chapter 8, outlines ways in which this happens.

The United States Constitution requires compensation to pay fair market value for acquired eminent domain property takings. As Sullivan (Chapter 3) notes, a number of incentives can be given to landowners to avoid US cases being taken to trial. Moreover, owners would be entitled to compensation for any interruption to use when the land holding is expropriated, even if the former use is later resumed.

There is similar protection in theory in the Australian Constitution, which requires compensation on just terms for compulsory acquisition of private property by the Australian Government. However, the Australian constitutional guarantee of just terms compensation does not apply to compulsory acquisition of private property by the state governments. This is replaced by legislation in all states that allows their governments to compulsorily acquire property and for compensation to be paid. The New South Wales state legislation provides for acquisition on just terms, with compensation to be at not less than market value. In the case of strata scheme terminations in New South Wales, implementation of the just terms criterion is made by the owners' corporation. The Land and Environment Court determines disputes about compensation by applying the 'just terms' criterion under the state's compulsory acquisition legislation.

The constitutional and legislative basis of compulsory acquisition and related negotiation and appeal rights

The legal basis of compulsory acquisition for densification, and associated negotiation and appeal rights, varies considerably from country to country. Perhaps the most differentiating feature is whether the powers to carry out such acquisition are constitutionally available or whether specific authorising legislation is required, with both avenues being used in some cases. Another distinction rests on the extent to which acquisition and negotiation/appeal procedures are closely prescribed by laws and regulations (as in Germany), or conversely the extent to which there is room for judicial or bureaucratic discretion.

The United States is the clearest example of compulsory acquisition (eminent domain) power being constitutionally based. This power is viewed 'as an attribute of sovereignty' as Sullivan puts it in Chapter 3, and thus does not require special legislation. In addition, the Fifth Amendment of the Constitution specifically allows expropriation of private land for public use, as noted above. As Sullivan also notes, special legislation may nevertheless be used to set limits to compulsory acquisition. Comprehensive planning policies under state statutes may control compulsory acquisition actions. An obligation for the government or public authority to negotiate with land owners may exist, depending on the state concerned. But, as Sullivan says, since eminent domain is constitutionally based, the opportunity for land owners to mount legal challenges is always there.

Germany is an example in which expropriation of property for densification and urban development generally is based on both constitutional powers and on ensuing legislation and regulations. Such expropriation is allowed under the German Constitution. It is specifically regulated under the Federal Building Code, which sets out the conditions under which compulsory acquisition can be effected. It includes a requirement that a serious attempt has first been made to acquire the land on acceptable terms. Once the application for expropriation has been made to the relevant authority (usually the state government), property owners are given the opportunity to make representations. All questions concerning the expropriation and compensation are discussed at the hearing held by the expropriation authority, which acts as an intermediary between the parties in an attempt to achieve agreement. If an agreement is reached, this is certified by the authority. If there is agreement on the change of ownership but not on the amount of compensation, the authority continues negotiations to reach an agreed value. If there is still no agreement, the authority determines the amount of compensation based on values calculated by a committee of experts. This value can be contested in a regional court by property owners.

The Netherlands has a somewhat simplified version of the German procedures. Under Netherlands legislation, compulsory acquisition is only allowed under strict conditions. It must be preceded by adequate voluntary negotiations to attempt to achieve agreement. If the land owner does not accept the government's final offer, the matter is taken to court. Appeals by either party against the decisions of lower courts can be taken to higher courts.

United Kingdom provisions are similar, although the relevant legislation and statutes are more wide-ranging though perhaps less transparent. The UK does not have a written constitution, and compulsory acquisition is based solely on statute/legislation. A key statute here is the *Local Government Act 1972*, which grants powers of compulsory acquisition for any purpose authorised by (other) legislation as well as for public benefits to an area. There is an explicit obligation on local authorities, set out in formal government guidance, to negotiate with property owners before compulsory acquisition powers are used. While there is no statutory obligation to demonstrate good faith in undertaking negotiations, acquiring authorities must take 'reasonable steps' to reach agreement. Government guidance indicates that compulsory purchase should be a last resort that is only used where there is a compelling public interest. Objections to compulsory purchase orders can be made on several grounds including those relating to the European Convention on Human Rights. Where objections remain after negotiations, legislation provides for the government to hold a local public inquiry using statutory rules of procedure. Objections relating to the amount of compensation can be challenged in the High Court.

Compulsory acquisition authorisation and procedures in Spain are prescribed by legislation. In the case of the most common form of urban redevelopment, land readjustment, acquisition legislation and regulations are quite complex in terms of land owner rights and obligations. Compulsory acquisition statutes involving outright expropriation rather than land readjustment involve simpler procedures and lesser rights for land owners, but such expropriation is only applied in special cases. Details of Spanish land readjustment legislation and regulations are set out in Chapter 7. In terms of land owner rights of objection and appeal in land readjustment cases, there are a number of stages. At the outset, the owners of at least half of a proposed readjustment area must agree to create a development organisation (junta) in which they organise the readjustment of their land. After approval of the project, land owners in some regions can request expropriation instead of participation in the project. Otherwise, land owners who are unwilling to be part of the project have their properties expropriated and are compulsorily incorporated into the project, with the same rights and obligations as those volunteering to be in the project. Public authorities have statutory powers to overrule land owners and apply compulsory readjustment if a majority of land owners do not join the junta or if there are delays in readjustment, and also to apply such adjustment from the beginning, without starting the voluntary process.

Singapore's Constitution excludes protection of property rights because it was argued that such protection would have hindered the need for the government to acquire sufficient land at a manageable cost to address poor living conditions. Instead, special legislation was enacted to allow the government to acquire land compulsorily. Appeals against the compensation awarded can be made to an appeals board, an administrative tribunal that has members with expertise in property valuation. Appeals against other aspects of compulsory acquisition have very limited scope, being restricted to *ultra vires* matters where the acquiring authority has acted outside its statutory powers, and acquisitions where the authority has acted in bad faith, as explained in Chapter 10.

In China, all land in cities is owned by the state. Households can own the residences on this land through self-build, for example, and it is these dwellings that can be subject to compulsory acquisition under a specific national legal regulation. More detailed regulations by district governments for specific projects can set out formulas for compensation, resettlement arrangements, scope for property owner participation in the process, and agencies for appeals. But in practice under current regulations, property owners cannot challenge urban development plans in which local governments can nominate areas for compulsory acquisition under the public interest criterion. Nor can they subsequently claim rights to remain in place. Rural land is legally owned by collectives. Again, it is the farmers' dwellings on that land that are subject to compulsory acquisition by local authorities. Local authorities exercise relatively unfettered powers under the Link Policy in the process of compulsory acquisition of rural homesteads. Rural households, like their urban counterparts, lack power to challenge acquisition and resettlement plans or claim their right to remain. Residents subject to compulsory acquisition orders in rural and urban areas can try and bargain for better resettlement outcomes and, as Chapter 8 notes, it is not uncommon for improved results to be obtained. But legal avenues for citizens to appeal against state decisions such as those involving compulsory acquisition are, as Zhang argues in Chapter 8, hindered by lack of judicial independence and the politicization of the legal system. In addition, the Administrative Litigation Law limits court authority to review legal conformity or procedural justice.

In Australia there is no constitutional restriction on compulsory acquisition or the power to dispossess by government. Thus this power is effected by legislation, as in the two Australian case studies (Chapters 11 and 12). In New South Wales, local government as well as the state government can compulsorily acquire land for public purposes: Chapter 12 deals with a case in which local government was found to have exceeded its compulsory acquisition powers. In New South Wales the scope for land owner negotiation and appeals lies principally within the just terms clauses of state compensation legislation.

National history of compulsory acquisition

While dispossession of private property by national rulers has a long history, the beginnings of formal compulsory acquisition powers date back about two centuries, with the UK initiating such powers to support the industrial revolution (see Chapter 4). The main purpose there involved property purchases to allow the expansion of the British rail system. The use of such powers in the UK expanded in the mid-nineteenth century to include other infrastructure such as reservoirs and sewerage systems, plus street widening. By the mid-twentieth century, compulsory purchase orders were used in association with new town planning legislation to facilitate slum clearance and redevelopment and town centre improvements. The same compulsory powers were also used from that period to facilitate the development of major transport infrastructure – motorways, ports and airports. Dutch legislation for compulsory acquisition dates back almost as far as that in the UK. The legislation originated in 1852 and has only been slightly modified since then.

Compulsory acquisition also has a significant history in Spain and the US. In Spain, expropriation for social/affordable housing started with the 1911 Act on 'Cheap Houses' (Chapter 7). In the US, such acquisition has a history dating back to the Fifth Amendment of the US Constitution, where it has been used for a wide range of purposes deemed to be of 'public use' that include various projects where the public use comprises economic development.

Singapore also has a particularly long history of compulsory acquisition arising from its colonial period. In the years after the treaty of 1824 ceding Singapore to Britain, the Colony reserved the right to resume private land for essential public developments (Chapter 10). From 1854, new leasehold titles empowered the Crown to resume possession for infrastructure. New procedures for allowing the Governor to acquire private properties for public purposes were legislated in 1920. Similar British-originating statutes also applied in Australia in the colonial period, although the need for compulsory acquisition of private property by the colonial governments was considerably reduced by their assumption of *terra nullius*, by which the government claimed ownership of all land from the time of inception of the colonies. After federation in 1901, the national and state governments all enacted legislation to allow compulsory acquisition of property. This has been used for a variety of purposes in the ensuing decades, providing for such purposes as slum clearance, rail and highway construction, and open space provision.

In other countries surveyed in this book, the first significant use of compulsory purchase powers has been for urban redevelopment (including associated densification) in the post-World War II period – early in this period in Germany for example, and later in China.

Restrictions on types of densification allowed under compulsory acquisition

No country discussed in this book has restrictions on the type of densification allowed by compulsory purchase powers, as long as the various public interest/use/benefit criteria under such powers are met. There are indirect restrictions, however, arising from requirements for densified development that results from compulsory acquisition to conform to relevant statutory land use plans. In countries such as the Netherlands, Germany, Spain and Canada (Quebec), for example, the main implicit or explicit role of compulsory acquisition is to help realise the intended outcomes of such land use plans.

Obligations to provide alternative housing

In several countries, compensation for compulsory acquisition involves the provision of alternative housing to replace that taken in the acquisition process. In the UK, while there is a duty by the acquiring authority to rehouse residents, there is a legal difference in alternative housing provision between owners/leaseholders and tenants of social housing (that is, dwellings provided by local government or non-profit housing associations). For social housing tenants, there is a legal duty to provide alternative accommodation, whereas for owners/leaseholders, the standard compensation rules apply (which include home loss payments). In London, the policy is that social housing tenants should be offered a right to return to the regenerated estate where the original dwellings had been compulsory acquired.

In China, resettlement of displaced urban or rural residents to new urban sites is planned by the district governments as part of the overall acquisition procedure. Dispossessed residents have to pay for their new dwellings out of compensation received for their old housing. However, the amount of compensation is frequently much less than the cost of the new dwellings. It is not known from Chapters 8 and 9 how resettled residents make up the difference in order to purchase the new dwellings.

Broadly similar arrangements apply in Singapore. The government offers displaced residents new accommodation elsewhere which is usually superior to their old residences (if the option of selling their old apartment on the open market instead of receiving compensation at existing market rates is not taken, as in the Rochor case study in Chapter 10). Public housing that is offered to resettled residents may be offered at subsidised prices.

The Spanish situation involving land readjustment schemes is distinctive in that residents whose land is acquisitioned as part of a scheme receive a new land parcel in the completed scheme. Rising land prices have meant that the value of the new parcels is considerably greater than the value of the original parcels. This still leaves residents the challenge of arranging their finances to build or buy a new dwelling, if the scheme has required their old dwelling to be demolished.

Other methods used to bring about densification

Compulsory acquisition is rarely the only way that governments have to achieve greater urban densities. Chapters in this book include some discussion of other methods that might be available, without purporting to give fully comprehensive analyses of such alternative methods. The central question arising from this concerns why compulsory acquisition is the chosen means of bringing about densification. While definitive answers are beyond the scope of this book, it is apparent that – as the discussion on practice in the Netherlands (Chapter 2) suggests – compulsory acquisition will be used as a last resort if other effective methods are available. This is because the procedure is complex and often costly, particularly in terms of procedures necessary to ensure just and fair processes. It is also because of the political opprobrium that attaches to the imposition of compulsory acquisition.

In the Netherlands there is a clear hierarchy of instruments that can be used to implement urban plans incorporating densification. Chapter 2 sets out this hierarchy from easiest and most preferred to least easy and least favoured, namely: voluntary land acquisition, voluntary land readjustment, pre-emption rights (which give municipalities the first right to buy a property), land consolidation and compulsory acquisition. As Holtslag-Broekhof, Hartmann and Spit note in Chapter 2, in practice the threat of possible compulsory acquisition can induce land owners to sell their land voluntarily, and effect that is also apparent in other national practices such as that of Spain. Voluntary buy-back policies are also seen as an alternative to compulsory acquisition in the UK.

In Germany, expropriation can be avoided by reallocation of property or (like the Netherlands) pre-emption. An alternative to the compulsory acquisition mechanism for achieving urban densification that is being considered at official level in Germany is the so-called 'infill development measure'. This would first involve the drawing up of a statute to determine the land use potential of a sub-area of a municipality. This would be followed by the imposition of building obligations with nominated deadlines on under-developed sites. Failure to meet the deadline would allow the municipality to offer a market price for the site, and if the owner rejects this, the possibility of expropriating the property exists as a last resort.

Australia has achieved significant urban densification with relatively little use of compulsory acquisition thus far (except for supporting transport infrastructure). One piece of existing legislation in New South Wales that has been used successfully is the *Growth Centres (Development Corporations) Act 1974 (NSW)* which permits a government development corporation to buy land in nominated urban areas (Chapter 11). Overall, however, the state government has preferred to achieve renewal and associated densification through the operation of private property owners in the market, using increased density limits in land use plans. The strata termination legislation that allows compulsory sale of apartments in older buildings effectively provides another incentive for developers to increase densities via the market. Similar legislation in Singapore (Chapter 10) provides an alternative there to redevelopment via compulsory acquisition.

Concluding thoughts

The great variety of approaches to using compulsory acquisition for densification that the chapters in this book have demonstrated suggests caution in extrapolating generalisations. With this in mind, a few concluding comments are offered.

A trend that is apparent across a number of countries is a shift toward more exceptional use of compulsory acquisition for urban redevelopment and/or more favourable treatment of land owners. In part this is because the intensity of mid-twentieth century housing challenges that necessitated a strong government response in countries such as the Netherlands and Singapore has now diminished somewhat (although the use of compulsory acquisition for urban infrastructure projects in Singapore continues apace). The need to use compulsory acquisition powers to redevelop urban slums in Australia as happened in the post-war period is no longer there with the clearance or gentrification of former slum areas. A backdrop to this trend is the rise in consumer rights and in participatory planning in western countries over the last few decades. But in China, the use of compulsory acquisition seems to have increased in recent years as urbanisation pressures mount and local authorities seek ways to finance their budgets.

Associated with this is a relationship between the need for urban redevelopment/densification and the strength of compulsory acquisition powers. Countries with the strongest expropriation powers or usage also tend to be those where pressures for redevelopment are greatest. China is the major example. There was a similar imperative in Singapore's early years of independence, but several measures which reduced compensation to property owners have now been replaced as housing and living standards have risen. The housing boom from the late twentieth century in Spain coupled with a relatively rundown housing stock has been associated with considerable use of compulsory acquisition, but implemented in a way that allowed property owners to share the fruits of the boom. Australia's big cities now have some of the most rapid population increases of any in the OECD, and resulting development pressures have helped to bring about two new compulsory acquisition statutes (Chapters 11 and 12) that have an absence of public benefit provisions.

Lastly, implications of the degree of discretion that is allowed in implementing compulsory acquisition deserve comment. Discretion in terms of the public benefits required from compulsory acquisition, through case-by-case court determination as in the US, or through lack of specification in legislation, as in Singapore and Australia, means that overall assessment of whether the benefits of compulsory acquisition outweigh the costs in those countries is very difficult. Benefits and costs may vary considerably from one case to the next: legal costs, for example, will go up or down depending on how idiosyncratic each case is and the extent of associated case law. But discretionary statutes allow wider freedom to argue points of justice and fairness. Conversely, where statutes and regulations reduce discretion, as in Germany and Spain, expected benefits and costs are more tangible. In those cases, this can serve to reduce the use of compulsory acquisition because procedural costs are very clear (Germany) or increase its use because property owners can base expectations of gain on a transparent procedure to value future financial costs and benefits (Spain). In the end, then, the question of whether the benefits of compulsory acquisition for densification outweigh the costs can only be answered by reference to the relevant legal statutes and related context of each country.

Index

Page numbers in **bold** denote tables, those in *italics* denote figures.

For Product Safety Concerns and Information please contact our EU representative GPSR@taylorandfrancis.com
Taylor & Francis Verlag GmbH, Kaufingerstraße 24, 80331 München, Germany

www.ingramcontent.com/pod-product-compliance
Ingram Content Group UK Ltd.
Pitfield, Milton Keynes, MK11 3LW, UK
UKHW011012180926
475701UK00004B/704
9781032095080